# SCIENTISTS IN THE CLASSROOM

## THE COLD WAR RECONSTRUCTION OF AMERICAN SCIENCE EDUCATION

*John L. Rudolph*

palgrave

First published 2002 by PALGRAVE™
175 Fifth Avenue, New York, N.Y. 10010 and
Houndmills, Basingstoke, Hampshire RG21 6XS.
Companies and representatives throughout the world.

PALGRAVE is the new global publishing imprint of St. Martin's Press LLC Scholarly and Reference Division and Palgrave Publishers Ltd. (formerly Macmillan Press Ltd.).

ISBN 0–312–29501–4 hardback
ISBN 0–312–29571–5 paperback

**Library of Congress Cataloging-in-Publication Data**
Rudolph, John L., 1964-
Scientists in the classroom : the cold war reconstruction of American science education
/ John L. Rudolph.
     p.   cm.
   Includes bibliographical references and index.
   ISBN 0–312–29501–4 (hardbound)—ISBN 0–312–29571–5 (pbk.)
   1. Science—study and teaching—United States—History—20th century.
I. Title.

Q183.3 A1 R83   2002
507'.1'0730904—dc21

                                                                    2001028508

A catalogue record for this book is available from the British Library.

Design by Letra Libre, Inc.

First edition: May 2002
10   9   8   7   6   5   4   3   2   1

Printed in the United States of America.

PRAISE FOR JOHN L. RUDOLPH'S
*SCIENTISTS IN THE CLASSROOM:*

" . . . absorbing . . . I wish everyone engaged in science curriculum development and those who study the history of curriculum and the history of science would read this book. It provides fresh insights into a complex phenomenon."

—Angelo Collins, Executive Director,
Knowles Science Teaching Foundation

*"Scientists in the Classroom* is a stunning revision of our understanding of educational policy making and public school reform after World War II. Eschewing conspiratorial notions of a corrupt alliance between scientists and the military-industrial complex, John L. Rudolph deftly uncovers the motivations and influence of the physicists and biologists who reshaped the high school science curriculum during the Cold War era. And he sets this fascinating story on a grand stage, populated with cold warriors, politicians, and reform-minded academics, amid an array of competing interest groups during the 1950s and 1960s.

We live in an age when the role of the federal government in educational reform in general remains contested, and when national security concerns in particular are preeminent and no less controversial. *Scientists in the Classroom* imaginatively recreates how the post-war generation of scientists, educators, and federal policy makers brought the Cold War into the classroom, with often unanticipated consequences."

—William J. Reese,
Professor of Educational Policy Studies & of History,
University of Wisconsin-Madison

*This book is dedicated
with love and appreciation
to Jenifer, Audrey, and Lydia*

# CONTENTS

# PREFACE

In the course of describing this project to various individuals over the years, the point has been raised more than once that it's all well and good to examine the intentions of education reformers and the materials they produced, but, in the end, these things tell you very little about what teachers did with those materials or what students actually learned when they encountered them. The teaching that takes place in schools inevitably varies from what reformers envision. I have been reminded that this was indeed the case for the science curriculum materials of the 1960s, which were commonly viewed as too difficult for typical students and thus were only partially implemented by teachers. All of this is undoubtedly true (though I would argue that fundamental changes took place nonetheless in what students learned and the way they were taught as a result of the scientists' efforts). What I hope is clear in the pages that follow, and what I emphasize here at the outset, is that this book is not about teaching and learning in the day-to-day sense, nor is it about the ultimate success or failure of the reform efforts described (another often-asked question). The focus of this project has been, rather, on origins and intents—on the question of how and why scientists became involved in pre-college education during this period in American history and what they did once they were in a position to exert their influence.

I concede that by treating these reforms from their beginnings just up to their implementation presents only one part of a much larger historical picture. But this part, it seems to me, is worth examining for a number of reasons. First, the connection to student learning should not be written off so quickly. Although educational programs are without question modified in the course of their implementation at the local level, the material products that accompany them (the textbook, visual aids, laboratory exercises, apparatus, and so on) remain the primary resources teachers draw on in their work. Thus, they undeniably play an important, if not the only role in determining what it is that students learn. It makes sense then to understand what those materials are and what they were designed to accomplish. In writing this book, however, I have deliberately fixed my attention on issues that reach far beyond those related to student learning. By examining the construction of school science in Cold War America—how it embodied tacit social relationships and the manner in which it was shaped by external influences and interests—we can learn a great deal about the perceived role of science in society at this point in history, about the malleability of the concept of science itself, and about how the institution of schooling and the science curriculum have been used (however successfully) as a tool to mediate the relationship between science and the citizen. It becomes evident, from this broader perspective, that this small part of the picture is really quite rich in historical and sociological insight.

Given my focus on the scientists in this reform movement, the story may at times seem one-sided. Throughout the book, I have tried to be sensitive to the other interest groups (professional educators, teachers, and so on) that had a stake in the changes taking place. But their voices are infrequently heard in these pages. This is partly a consequence of my choice of subject. But it is also due to the fact that, although superficial efforts were made to include teachers and education professionals in the work being done, most of these individuals were deliberately and effectively relegated to the sidelines. The few initiatives that were undertaken by educators almost uniformly adopted the language and methods of the scientists who had taken center stage directing educational policy during these years. Objections were raised that educators' voices were not being heard as the reform movement gathered momentum, but few offered any viable alternatives to the government-embraced educational vision of the scientists. The result is a narrative that reflects their almost singular approach to reform.

What should be evident from these prefatory remarks is that productive lines of research remain to be cultivated, be they related to questions of classroom implementation, to accounts of how other groups adapted to these educational changes, or to how nonscience subjects were similarly transformed. Years ago in graduate school, I was surprised to learn that these innovative, federally-supported reform efforts had received so little scrutiny from either historians of education or historians of science. That was, of course, to my advantage then. But more work needs to be done, and I hope that this book might provide something of a stimulus to further studies in the history of science education generally, as well as work aimed at a better understanding of American education in this particular Cold War period.

Financial support for this project has come in many forms. The early stages of my archival research were made possible by a National Science Foundation grant (SES-9632778). A Tashia F. Morgridge Fellowship at the

University of Wisconsin-Madison provided me with resources to conduct additional research and, more importantly, much needed time to devote to writing. And the Wisconsin Alumni Research Foundation made available funds that allowed me to put the finishing touches on the final product. I gratefully acknowledge all these benefactors.

There are a number of individuals I would like to thank for their help with this project. The many archivists and librarians at the institutions I have had the pleasure to visit come first to mind. Of these, I would like to mention in particular Dwight Strandberg at the Eisenhower Presidential Library and Janice Goldblum at the National Academy of Sciences Archives. Numerous people have read and commented on portions of this manuscript, these include Michael Apple, Peter Hewson, David Kaiser, Adam Nelson, Ron Numbers, Bill Reese, and Sam Schweber. Their comments and suggestions, though not all incorporated into the manuscript, have been greatly appreciated. Thanks also goes to Donna Schleicher and Simone Schweber, and to my editors Michael Flamini and Amanda Johnson for their advice and encouragement. Herbert Kliebard and Lynn Nyhart, deserve special mention for seeing this project through from the beginning and for their support throughout. Jim Stewart was also there from the start and to this day remains a trusted advisor, colleague, and friend.

Finally, this book would not have been possible without the support and understanding of my family. For their help in countless ways, I thank my parents Donald and Connie Carter, Susan and Dennis Cotey, and Jerry and Mary Ann Maus. In addition, Vincent and Sandra Rudolph were particularly helpful in providing me with a home away from home on my many research trips to the Boston area, which made my work there seem much less like work. And, as much as I might try, it seems impossible to appropriately express my deep appreciation for my wife Jen for her love and understanding (and numerous readings of these chapters) all the years we have been together. I thank her especially for the wonderful home she has created for our daughters, Audrey and Lydia, who make this all worthwhile.

*John L. Rudolph*
*Madison, Wisconsin*
*December 2001*

# INTRODUCTION

E arly in 1960, the Educational Policies Commission of the National Education Association (NEA) circulated a draft policy document that sought to clarify for the nation the necessary function of the public schools. The manuscript, entitled "The Controlling Purposes of American Education," was occasioned by the rapid changes that were taking place in all facets of society. It began by laying out the fundamental assumptions that needed to be considered in reconstituting education in the United States. Foremost among these was the new social role that science had assumed. "The pursuit and use of scientific knowledge are the most fundamental forces changing the world today"; "With every day [science] becomes bigger and more important"; "Faith in its validity increases. . . . It seems to carry the promise of the future," the authors wrote. There was no less faith in the social role of public education. Given the importance of science, the Commission deemed it imperative that "schools and colleges . . . provide educational experiences which will bring to all the people a general understanding of scientific knowledge, awareness of the process of rational inquiry, and dedication to the proposition that expansion of knowledge in this area is the price of survival." The country's task was, above all else, "to permeate the educational environment with the empirical, rational spirit."[1]

What is so striking about the ideas expressed in this document is how foreign they seem to our current sensibilities a mere 40 or so years later. Today science is hardly viewed with universal esteem, as the sole path to social progress. And public education, far from being embraced as an institution for the common good, is continually under attack from interest groups seeking everything from privatization to complete elimination. The union of the two—public education directed toward the widespread assimilation of scientific thinking—is clearly at odds with current conceptions of the role of schooling in a pluralist society and even with the stated views of the Commission itself only a little over a decade earlier.[2] Thus, the interesting historical task here is trying to make sense of what seems to have been a "golden age" of science education in the United States as the 1950s gave way to the 1960s. Why exactly was the notion of rational inquiry as manifested in the practice of science held in such high regard? Why were the public schools seen as central to its dissemination? And how exactly would this far-reaching goal be accomplished? Unallayed faith, optimism, and progress were hardly the most notable characteristics of 1950s American culture. The reference by the Commission to science education being the "price of survival" indicates a darker undercurrent of anxiety and turmoil that mixed uneasily with the companion visions of hope and progress. All of these elements, as we will see, are central to understanding the essence of this "golden age."

This declaration of the purposes of education by the NEA, a group not traditionally aligned with the interests of scientists, represents, more than anything, the triumph of the nation's scientific elite in generating a public-policy consensus in the early 1960s regarding the primary goal of American education, a point that raises the added question of why the nation's scientists felt the need to involve themselves in the schools in the first place. In examining the history of science in this period, one finds that education, particularly pre-college education, was but one component of a broader effort on the part of leading members of the scientific community to reshape the social and cultural environment in which they worked. It included forays into public relations and advertising, science writing, television, and films.[3] Despite the variety of media outlets scientists sought to exploit, education remained the central plank in their movement, and the most tangible manifestation of their efforts can be found in the science curriculum reform projects that made their appearance in the mid-1950s.

The curriculum reform movement of this era is widely identified with a host of project acronyms—PSSC, CHEM Study, BSCS, SMSG, CBA, and IPS, among numerous others. The new programs spanned nearly all the science disciplines, from physics to the earth sciences, and generated a host of new instructional materials including elaborate film presentations, innovative laboratory apparatus, case studies, supplementary readings, and, of course, textbooks. By 1977, over 64 percent of all school districts had adopted at least one program from the initial set, and at one point it was estimated that 19 million students were enrolled in courses utilizing the new programs. The magnitude of this collective undertaking in terms of scope and funding was indeed phenomenal.[4]

The defining characteristic of these projects was the prominent role of the federal government and a handful of elite academic research scientists as the architects of change. In a unique collaborative effort, the United States Congress appropriated funds, the National Science Foundation

(NSF) provided guidance, and scientists of various sorts worked together to realize their vision of what science education should be. In a nation where the school curriculum had long been shaped by the diffuse competing interests of various professional organizations, publishers, national committees, task forces, and the tens of thousands of local school boards nationwide, the centralized approach adopted by the country's top scientists after World War II was a bold departure that had a significant long-term impact on national education policy. As curriculum historian Herbert Kliebard remarked, after the war "the way in which the curriculum of American schools was determined was never quite the same."[5]

Although scientists had moved swiftly into the field of pre-college science education in the late 1950s, their active engagement there was only temporary, lasting little more than two decades. By 1982, following a rather explosive public controversy over an NSF-sponsored social studies curriculum in the mid-1970s, federal funding for curriculum reform had all but dried up. And while NSF was able to reestablish some semblance of its curriculum development program years later, it was never again able to match the sheer size or sweep of the earlier effort, nor did it enjoy the level of leadership from research scientists that marked those initial curriculum experiments. Thus, as quickly as the reform movement had burst onto the scene, it seemed to have receded into the shadows, replaced by the social and economic educational priorities of the 1970s.[6]

The broader purpose of this book is twofold. In recounting the history of these science education reform efforts, I hope to understand more about how scientists, and their allies within the federal government, sought to use education as a tool to manage the relationship between science and the lay public. At the same time, I expect to provide some insight into the broader question of how educational policy and classroom materials are produced in the United States. Naturally the intersection of the scientific community and public education allows for an examination of each in light of the other. The details of what I am calling the Cold War reconstruction of science education tell on both the inner workings of the scientific community during a crucial period of institutional growth and political struggle as well as on the institutions and practices of schooling in the United States. At the intersection of these two institutional domains, at least in this instance, lies the school science curriculum.

An understanding of the events surrounding this episode of reform, however, has more than just historical value. Though the movement clearly failed to achieve the widespread changes it was designed to bring about, it has had a significant and lasting impact within the field of education. The new curriculum projects, whether they influenced students or not, succeeded in establishing enduring norms of content and instruction that continue to shape school science to this day. Much of the language, practices, and expectations for science education have been derived from this movement. The rehabilitation of subject matter, elevation of the instructional role of the laboratory, utilization of innovative instructional media, and particularly the emphasis on discipline-centered inquiry and explicit attention to the nature of science—all commonplaces in the education literature today—can be traced to these early NSF curriculum projects.[7]

To appreciate the nature of the educational materials the scientists produced and how they came to produce them, though, one must understand the historical context in which they worked. In the pages that follow, I examine the social and political conditions that enabled these NSF-backed scientists to gain the authority to prescribe the substance of the high school science curriculum at a time when public schools were particularly sensitive to issues of federal encroachment in local affairs, and I trace how the material products of the various curriculum projects were conditioned by those social and political forces in play at the time. In moving from the disciplinary practice of a given science to the construction of its corresponding school subject, there were multiple points at which historical circumstance and social forces no doubt impinged. I intend to highlight what I argue are the most important for understanding the changes that took place.

Part of the impetus for reform came in opposition to the life-adjustment curricular ideology that dominated national education meetings, administrative councils, and classrooms in the years after the war. The life-adjustment curriculum, which had its roots in the progressive education movement, was the epitome of functional schooling, in which academic subject matter was marginalized in favor of courses designed to meet the immediate social, personal, and vocational needs of the student. While academics and scientists charged that it emphasized methods of instruction over content and promoted emotional adjustment at the expense of learning, it also provided a convenient target for attacks that had deeper ideological roots—roots that were laid bare by the intensification of the Cold War.

The course of reform during this period in history cannot be understood without considering the profound influence of this grinding conflict. The adversarial relationship between the United States and the Soviet Union in the 1950s cast a long shadow that touched all facets of national life. Along with the international military tensions came an ideological cloud that settled across the American landscape, which appeared to distort nearly all public discourse. The debates over the life-adjustment curriculum were, from the start, polarized along these stark ideological lines. But as important as the domestic politics of education were, more far-reaching Cold War forces conspired against the status quo. Few segments of society were immune to their influence, and the scientific community felt them more than most. The everyday context in which scientists lived and worked during these tense years is central to understanding their involvement in American schooling.

After witnessing the remarkable contributions scientists had made during World War II, the federal government worked to incorporate scientific expertise permanently into the national security arsenal. But even as they were drawn in, scientists struggled to set the terms of their service. They sought to maintain the high levels of funding to which they had grown accustomed during the war, yet at the same time worked to safeguard their autonomy from the government control that inevitably came with public support. The public's understanding of science was seen as crucial to this effort. Yet as the military significance of science became magnified during the Cold War, there seemed to be a corresponding magnification in public misunderstanding of what scientists did. It was this misunderstanding that their educational efforts were designed to correct.

But there is more to consider than simply the conscious desire of scientists to recast the perception of scientific work for the sake of their own advancement. Another key element of curriculum reform included bringing what many scientists viewed as the antiquated content of existing school science into line with modern scientific knowledge, to present science "in the framework of 1956 instead of 1896," as one project director explained. Here more subtle forces were at work.[8] The science of the 1950s was indeed different than that of the previous century, different even than that of the previous decades. World War II, a societal watershed in so many ways, had multiply transformed the nature of scientific inquiry. The Cold War solidified those changes. Despite their enduring concern with professional autonomy, scientists had been conditioned by their wartime success to accept new organizational patterns of research. Large-scale, goal-oriented science (the legacy of the "crash" programs epitomized by the Manhattan Project) replaced the diffuse work of solitary investigators, and the emerging political economy of science privileged manipulation and control of the natural world over all other goals. In modernizing the curriculum, scientists drew on these techniques and research practices with which they had become so adept. The very structure of the school science they produced, as we will see, inevitably bore the imprint of these fundamental, war-induced changes in scientific knowledge and practice.

From the numerous science curriculum projects initiated during this era, I have chosen two to tell this story: The Physical Science Study Committee (PSSC), led by physicist Jerrold Zacharias at the Massachusetts Institute of Technology (MIT), and the Biological Sciences Curriculum

Study (BSCS), jointly headed by Bentley Glass and Arnold Grobman at the University of Colorado in Boulder. PSSC, which got its start in 1956, was the first of the federally-funded curriculum projects and unquestionably provided the model for those that followed. BSCS, which began two years later, provides an interesting contrast to the physics project. The work of the biologists illustrates not only the pervasiveness of the curricular vision advanced by the physical scientists, but also the particular interests of a discipline that was largely marginalized in the postwar period. On equal footing with PSSC and BSCS in this historical study is the National Science Foundation, which contributed more than mere funding for curriculum reform. NSF, as both a representative body of the nation's scientific elite and an official agency of the federal government, played a central role in effectively brokering the entrance of scientists into the field of pre-college education.

This book is divided roughly into two main parts. The first, which spans the first three chapters, serves two functions. It recounts the circumstances and events that led NSF to initiate its comprehensive program of curriculum development, which brought the country's top research scientists into the business of educational materials production. It also furnishes the broader historical context necessary to understand both the scientists' motivation to engage in such work and the resulting form those curricular materials assumed.

Chapter 1 sets the stage, describing the deepening Cold War and the rising domestic political tensions it brought. It was in this climate that the school curriculum became a point of contention. I mark this beginning in 1949 with the onset of the Red Scare and the first high-profile attacks on the life-adjustment curriculum, which began with the publication of Mortimer Smith's book, *And Madly Teach*. Illinois historian Arthur Bestor, however, soon became the intellectual leader of a key faction of the school critics. Bestor and, to a lesser extent, Smith represented the growing dissatisfaction of a group I have chosen to call the academic traditionalists, who, in their scathing attacks on the reigning curriculum type, served to throw open the question of what sort of education would best meet the public interest. One of the most significant aspects of this debate was the highly-charged ideological rhetoric used to frame the arguments of the participants.

In the next two chapters, I provide an account of the evolving place of science and the scientist in American culture during the postwar period. Specifically, I examine the scientific community's changing relationship with the federal government and the scientists' views of how public confusion over the nature of science affected that relationship. I then illustrate how these historical circumstances and existing public perceptions facilitated the ascension of the scientists to their central role in education reform. Many scientists believed that the school curriculum could function as an effective vehicle through which they might more effectively remake the public attitudes toward science. Here I argue that it was the technological contest with the Soviet Union and the seemingly nonideological nature of science itself that, in the context of the earlier curriculum debates, allowed these individuals to move into the field of education with few objections from either Congress or the general public.

The last four chapters constitute the second part of the book, which deals specifically with the development of curriculum materials by PSSC and BSCS. My intent here is to show how the political and historical circumstances set forth in the first few chapters came to shape both the manner in which these materials were produced as well as the final form they assumed. Chapter 4 relates the carryover of the new techniques and modes of organization that characterized the large-scale, analytical approach to research that emerged from the war to the development of the physics and biology curricula at MIT and Colorado.

Chapters 5 through 7 focus on the activities and deliberations of the PSSC and BSCS projects individually. I demonstrate how the specific content and instructional activities of each were explicitly designed to convey a view of science consonant with the interests of the scientific establishment, and I begin to untangle how the broader changes in postwar scientific research became reified, though less intentionally, in the substance of the two projects as well. One can see how the scientists' desire to maintain the political autonomy of their work became subverted by the military imperatives that had made their way into the practice of physics and biology. What surfaces from the confluence of scientific and national security interests in the curriculum is a discipline-centered view of science that showcased process as well as content. Both, however, were unquestionably skewed toward intervention in and control of the natural world—goals valued by the military patrons of scientific research. The more humanistic goals of the scientists were, in the end, compromised by the technocratic orientation science assumed in the postwar period.

Though nearly all the curriculum projects have passed into disuse, tagged as too difficult, elitist, or themselves now outdated, their dramatic appearance in the 1960s produced a profound shift in our view of science education. These projects grew out of existing classroom practices, international and domestic political tensions, wartime technologies (hot and cold), disciplinary rivalries, and above all the professional desires of the American scientific community. All were embedded, in multifarious ways, in the very matrix of the new school science. Though the image of science that the scientists sought to project fell somewhat distorted on the American public, its effects were nevertheless lasting. It has endured like an afterimage burned on the retina long after the eyes have been closed, an image that has continued to shape the conscious perceptions of what science is and what science education should be. This book is an attempt to understand the circumstances of its formation.

# CHAPTER 1

# IDEOLOGY AND EDUCATION

As the hostilities of World War II drew to a close in the middle of the twentieth century, the war-weary nation turned its attention to long-neglected domestic concerns. Americans entered the new peace fully expecting to realize their dreams of living the good life they had fought to secure. Pent-up savings, along with government housing and education programs for returning GIs fueled a postwar boom that lasted throughout the 1950s. New suburban communities sprang up across the country and were quickly filled by "the most amazing social trend of the postwar era"—the baby boom.[1] The growing population of children soon made its way into the nation's classrooms, drawing public attention as never before to the schools. But, while parents busied themselves with bake sales, PTA meetings, and school plays, international events generated an underlying anxiety that touched the daily affairs of nearly all Americans. Since the end of the war, the nation had only begun to learn to live with the atomic bombs that had burst onto the scene in the skies over Japan. The existence of nuclear weapons, even in a peaceful world, was enough to make one pause and ponder the ultimate fate of humanity. The growing Cold War with the Soviet Union, however, brought home the possibility that another, more devastating war might actually take place.

In the immediate postwar years, U.S. and Soviet political leaders had begun to mark off the boundaries of a battlefield far from those on the European continent or in the Pacific. "The cause of freedom is being challenged throughout the world today by the forces of imperialistic communism," declared President Truman in the spring of 1950. The new conflict was one of ideas, a contest between the philosophical foundations of American-style democracy and Soviet communism. It was a battle of truth against propaganda in the eyes of many Americans, and the United States was fully engaged, according to Truman, in this "struggle . . . for the minds of men."[2] With such provocative rhetoric at the highest levels of government, it was probably inevitable that controversy would erupt over the role of the public schools—the institution primarily responsible for shaping the minds of the nation's most intellectually vulnerable citizens: the children. In the early 1950s the school curriculum, in particular, came under intense scrutiny and became an important ideological battleground on which partisan groups clashed as the nation's survival seemed to hang in the balance.

The early curriculum battles were intense and widespread, played out in the pages of popular magazines and best-selling books across America. Although a number of interest groups participated in these very public discussions regarding the proper aims of schooling, two groups emerged to set the terms of the debate. On the one side were the professional educators, who since the 1930s had secured the public school curriculum as part of their professional domain. Adhering to the progressive-education-type philosophy that favored practical knowledge over academic subject matter, these educators disseminated a curricular approach that stressed education for life-adjustment for all students in the years after the war. On the other side were the academic traditionalists, led by University of Illinois historian Arthur Bestor and his colleagues, who advocated just the opposite, a curriculum consisting of the long-established disciplinary subjects such as English, history, the sciences, and foreign languages.

The public criticism of the existing practical curriculum by those in the traditionalist camp was an important precursor to the scientists' entry into the field of education. It helped clear a path for their work in two ways: First, the highly-publicized attacks effectively redirected some of the public concern over the material problems of the schools (the growing building and facilities shortage) to their perceived instructional shortcomings, calling into question the adequacy of a school program that de-emphasized disciplinary knowledge at time when intellectual disciplines such as science and math were rapidly increasing in importance. Second, Bestor and his colleagues, by utilizing ideological criteria in their critique, served to legitimize the kinds of political attacks that had previously only existed at the disreputable fringes of public discourse. As a result, questions of democratic compatibility became increasingly relevant, and, in the political climate of the time, this meant educational programs that would ensure the survival of democracy in the face of the Soviet ideological challenge. The Cold War demanded a school curriculum that would provide the necessary intellectual rigor to compete internationally with the Russians and, at the same time, reinforce American democratic values. The life-adjustment program with its emphasis on the personal needs of students and socializing individuals to the larger group (which had collectivist overtones) appeared to fail on both counts.

Given these circumstances, and with the Cold War only growing in intensity, education reform seemed increasingly likely. Whatever emerged, however, would have to pass these stringent ideological and intellectual tests in order to be embraced by the public and welcomed into the nation's classrooms. The scientists, who largely remained on the sidelines during this period, found themselves nevertheless particularly well positioned in the years following the controversy to offer an educational program that would satisfy these requirements.

### Crisis in the Schools

With respect to the immediate problems that schools faced after the war, as Arthur Zilversmit noted, "the most important facts could be expressed in numbers, overwhelming numbers."[3] The baby boom made its presence felt and was one manifestation of what Elaine Tyler May described as a pronatalist ideology that came to characterize the postwar domestic period. Creating a safe, tranquil home life not only satisfied those yearning for a return to normalcy after the war, but also signaled a reaffirmation of American values in the face of the Cold War struggle with the Soviet Union. According to May, having children, in part, demonstrated "loyalty to national goals" and became an "expression of one's citizenship."[4] Thus, the family, in this interpretation, carried with it a good deal of political symbolism in an era fraught with ideological tension. At the very least, this newly realized focus on the family gave rise to nearly 4 million new children every year well into the 1960s.[5]

Children born during the war began flooding into the schools noticeably by 1948. By the fall of 1953, first-grade enrollment had increased by 34 percent with no relief in sight. Though high school enrollment remained manageable in the early 1950s, even declining in 1950, the demographic crest proceeded inexorably into the upper grades by the mid-fifties, and total enrollments were expected to peak in 1964. The educational infrastructure, woefully neglected during the Depression and subsequent war, was strained beyond its capacity. *U.S. News and World Report* estimated in 1953 that three quarters of a million classrooms would be needed to meet the rising demand the new students presented. Existing schools were in deplorable conditions, especially in the South and other rural areas of the country; fewer than half of all schools nationwide had indoor toilets. By 1955 students in the suburban communities, hit hardest by the baby boom, were attending classes in makeshift tents, local churches, and modified gymnasiums as school construction struggled to catch up. Textbooks and general supplies were also hard to come by; often children had to double up or do without.[6]

Further straining the integrity of schooling was the acute shortage of qualified teachers. Colleges and universities were only graduating a fraction of those needed, forcing districts to increase class size and, more troubling to some, hire unqualified applicants. Despite the progress made in increasing the education requirements of teachers over the past three decades, the circumstances of the time proved to be difficult to overcome. College students found little incentive to enter the teaching profession, and the few well-qualified teachers increasingly turned their backs on the classroom, citing low pay and prestige, as well as greater opportunity in the rapidly expanding postwar economy. Those left behind willing to tend the overcrowded classrooms were far from the intellectual leaders of the day. Horror stories of children being taught by cab drivers and busboys made their way into the news, first through a series of articles based on a six-month survey by *New York Times* education reporter Benjamin Fine in 1947. Thereafter the "school crisis special report" became a yearly fixture in the fall issues of the popular news magazines as the demographic bulge made its way up through the grades.[7]

As local communities set out to tackle their individual crises, few found any inexpensive options available to them. Per pupil costs had doubled in the decade after 1940 and the unavoidable school construction costs, estimated to reach upwards of ten billion dollars before the end of the 1950s for the country as a whole, placed an enormous burden on district finances.[8] Luring qualified teachers back into the classroom required increasing salaries as the competition for their services intensified.[9] In this crisis atmosphere, the National Education Association (NEA), the nation's largest teachers' organization, made a strong push for federal aid to schools. A perennial issue since the late 1800s, federal aid legislation seemed poised to make a breakthrough. Opposition had traditionally arisen to funding religious schools and to federal infringement on "local control," a euphemism used by legislators (particularly those in the South) to safeguard established systems of racially segregated schooling. However, with the school crisis nationwide, the support of President Truman (the first president to so favor general federal aid to schools), and the postwar prosperity of the time, the passage of an aid bill in some form seemed possible.

It was not to be. Controversy over aid to Catholic schools erupted and upset the favorable alignment in Congress. The moment was lost, and despite additional attempts under Truman and similar token efforts by the Eisenhower administration, no general federal aid bill was able to pass.[10]

This inaction by the federal government left the states and local districts alone to manage the school crisis. Conditions forced school boards and municipalities to raise taxes and push through bond issues to meet their needs—moves that sparked considerable controversy as state and local debt ballooned. The country had moved sharply to the right after the war politically; Republicans gained the majority in Congress in 1946 on promises of "a cutback in the role and spending of the government, and lower taxes"—issues that resonated strongly at the local level as well.[11] As districts moved to finance needed expansion, citizen groups sprang up to object to what they felt was excessive spending by school boards for lavish facilities and other educational frills. Critics soon extended their complaints beyond school finance to touch on subjects as varied as bureaucratic centralization and control, employee qualifications, and teaching methods.[12]

Local school battles broke out across the country between 1949 and 1953—a period referred to by many as the "great debate" in public education.[13] Criticism of public schooling was so overwhelming during these years that the NEA established the Committee to Preserve Free Education in its defense.[14] Given the explosive growth of school enrollments combined with the increased attention to family and child rearing by more highly educated parents in the early fifties, one would have expected the schools to come under rather intense scrutiny. The nature of the public criticism, though, was often long on rhetoric and short on specifics. Tax issues were least ambiguous. School funding, however, rarely exhausted the stock of objections critics raised.

## Subversion in the Classroom

Vocal opposition to tax increases soon invited dissatisfaction with the nature of the instruction taking place, and the turmoil created by the school building crisis and teacher shortages opened the door for reactionaries to take the lead. Some railed against the increasing secularization of the curriculum, others against its advocacy of internationalism. The United States had taken on a new role in an uncertain world. The public schools, reflecting the times, seemed to some parents to be ushering in a modernist, cosmopolitan culture very different from what they experienced growing up.[15] For more extreme critics, the culture reflected in the school curriculum was not only different, but also more disturbing, perfectly suited to the goals of communist expansion.

The threat to peace and prosperity from the Soviet Union, though not felt directly on the American continent, was nevertheless perceived as growing. American foreign policy during the Cold War operated on the assumption that the Soviet leadership was driven by a communist ideology that was inherently expansionistic, with world conquest as the only acceptable historical outcome. Many saw communism as "a monstrous and monolithic global conspiracy centered in Moscow and bent upon absolute world domination." To many, conflict with the Soviets seemed unavoidable.[16]

International events quickly lent credence to such fears. Communists took control of Czechoslovakia in February 1948 with Soviet help, and four months later the Soviet army sealed off Berlin, heightening international tensions as Britain and the United States airlifted supplies to the western sector of the city. The events that most shook the American psyche, though, occurred in the fall of 1949. On September 24, Washington announced that the Soviet Union had successfully exploded its own atomic bomb, thereby ending the short-lived American nuclear monopoly. Less than a week later the communists took control of mainland China.[17] The right-wing popular press wasted little time in fanning the flames of public alarm. Throughout 1950, readers were provided a steady photographic and textual diet of communist uprisings around the world alongside pieces questioning American resolve and military preparedness. If anything seemed clear in the closing months of 1949, it was the fact that communist power was indeed growing; American military power was no longer unrivaled.[18]

With China's fall to communism and the loss of the atomic weapons monopoly came the sense that the United States had begun to lose control over the course of international events. The seeming failures of 1949 prompted critics to look beyond existing economic and political structures for the cause of these challenges to American omnipotence. The only acceptable explanation for the undesirable turn of events, for them, was that the country must have been betrayed from within. The year 1949 thus marked a key turning point in American history: military intensification of the Cold War abroad made conflict with the Russians appear imminent, which helped usher in a sweeping Red Scare at home that was to permeate all facets of life well into the 1950s.[19]

A spate of espionage revelations provided the necessary ammunition to begin the purges. In January 1950 former State Department employee Alger Hiss was found guilty of perjury in a case involving the passing of government documents to the USSR. Less than a week later, Klaus Fuchs, a British scientist who had helped develop the atomic bomb, was arrested in Great Britain for passing information to the Soviets.[20] The invasion of South Korea by the North in June and the rapid U.S. involvement added significantly to the crisis atmosphere. Soon Joseph McCarthy would indelibly forge his name into the American memory as the crusade against communists and civil liberties began in earnest. Ideological tests were the order of the day as zealous citizens sought to smoke out subversives from sensitive positions in all levels of society.

The "Red Scare" provided a point of focus for a good deal of public anxiety, particularly with respect to the schools. Communist subversion seemed to provide the explanation many were looking for. With the war going on in Korea, and media attention on enemy techniques of brain-washing, propaganda, and political indoctrination (psychological tools to which children were viewed as particularly susceptible), it is not surprising that the tenor of school criticism turned toward its subversive potential. The Cold War, as historian Stephen Whitfield has documented, penetrated deeply into American culture: "Unable to strike directly at the Russians, the most vigilant patriots went after the scalps of their countrymen in-stead."[21] Public schools, so susceptible to the whims of the local community, were easy targets for those frustrated by the course of worldwide events.

Members of reactionary citizen groups spent little time worrying about intellectual justifications for their attacks on communism in the schools. For most, the fact that communism was un-American (though what that meant was far from clear) was more than enough reason to rid the schools of its influence. Some individuals in the academic community, however, did begin articulating an intellectual stand against communism. Among in-tellectuals, this grew into a consensus view of the Communist Party that became the basis for mainstream efforts to purge communists from all lev-els of the American political system as well as from the nation's schools and universities.[22]

The specific argument against communists in schools was pressed by liberal intellectuals seeking to preserve their privileged place in the acad-emy. With the resurgence of the Red Scare, these individuals (often viewed with suspicion in American society) sought political cover by giving up their communist colleagues in the name of academic freedom. Their argu-ment for doing so depended on two points. First was the definition of the Communist Party as a conspiratorial organization that employed unethical means to meet its political objectives. Second, they argued that member-ship in the party entailed a strict adherence to a political party line. Thus, on the first point, the secretive nature of Communist Party membership was patently unacceptable in institutions of higher education, places de-voted to the open and free exchange of ideas. The second point, though, was the more powerful and came to represent the defining characteristic separating communist ideology from other systems of thought. (It would prove later to bolster the public credibility of science.) According to this view, by joining the party one necessarily accepted the ideological system on which it was based and indicated his or her willingness to follow the party line in all matters, which included the indoctrination of students. In the eyes of liberal academics, Communist Party members had, as a result of their voluntary association, given up their commitment to free thought, the sine qua non of a university scholar.[23]

Sidney Hook, chair of the philosophy department at New York University, stridently explained this fundamental incompatibility between party membership and honest scholarship in a series of articles in 1949. Since the Communist Party required adherence to its ideological system, Hook argued, membership in the party makes a teacher intellectually dishonest by default. For in joining the party "he has signified his willingness to teach *according to directives received* and not in accordance with objective methods of searching for the truth." Even when the teacher disagrees with the party line on some points, he is to defer for the greater good of the party. It is precisely this "evaluation of what is important or unimportant in the light of a political objective," Hook went on, "that makes it impossible for him to exercise the free criticism he would engage in were he loyal to the principles of scientific inquiry." In contrast, the true scholar "is prepared to learn from anyone. . . . Doctrines are only valid or invalid in the light of objective evidence and logical inference."[24] Dedicated scholarship was not only anathema to communism, but, properly practiced, could reveal its flaws in the ideological conflict partisans were then waging.

This rationalization provided the semblance of legitimacy necessary to continue the expulsion of communist teachers already begun. The target, however, soon shifted. The power to indoctrinate, politicians profitably learned, was not limited to communist individuals. As they became scarce, an easier target was found in the books and materials used in schools, about which almost anything could be said without fear of rebuttal. In the spring of 1949, the House Un-American Activities Committee requested from 71 colleges and universities lists of textbooks used in all their courses. Many ignored the request; a few complied. All objected to Congress extending itself beyond its natural jurisdiction in this way, and, in the face of strong opposition, the committee abandoned the investigation.[25] Red hunters were to find greater success in the lower schools.

In a widely publicized case in Pasadena, for example, a local self-appointed watchdog group demanded that the school board justify on ideological grounds the curriculum it had approved for district use. The group wanted to know why, in one instance, a textbook directed students to consider the reasons for the failure of past democratic civilizations. Clearly, the group stated, such material was designed to "'sell' our children on the collapse of our way of life and substitution of collectivism." They demanded that the school board "determine immediately the politico-social aims of the present school administration," claiming it sanctioned a school program that fostered un-American attitudes in children.[26] In a broader attack in 1952, the *American Legion Magazine* published an article, widely circulated among reactionary groups, entitled "Your Child Is Their Target." In it the author lambasted the educational establishment for promoting so-called "progressive" curricula that indoctrinated children with "new social and political attitudes" for the purpose of "mold[ing] the welfare-socialist state." The extremist tone is evident in the opening paragraph of the article, which began, "Do you recall the parades of regimented children of Russia, the thousands of young communists massed in Red Square? . . . Have you ever asked yourself how did those children get that way?" "Indoctrination did it," was the author's confident reply.[27]

The effect that this ideological scrutiny had on the day-to-day teaching and learning that went on in the schools is difficult to gauge. For many it was simply a non-issue. Most schools were far from the seedy dens of socialist mind control that the national critics suggested. Historian Arthur Zilversmit has looked carefully at a number of schools and their curricular practices in this period and claims that community members on the whole were pleased with the curriculum teachers presented to students. A poll taken at the time in the Illinois counties Zilversmit studied showed that only two to ten percent of all parents were dissatisfied with the schools their children attended.[28] Though large numbers of schools across the country were left untouched by the extremes of both progressive education reforms and the attacks of reactionary citizen groups seeking to eradicate socialist indoctrination, the ideological quarrels of the day had an impact nonetheless. The great debate over the public schools was national in scope. Battles over the content of schooling made headlines throughout this period.[29] Reactionaries like Allen Zoll, who supplied the red hunters in Pasadena and others with anticommunist literature, became news makers almost overnight. For many outside observers and policy makers, the turmoil, though perhaps not reaching all the way into their own communities, was both real and pervasive.[30]

## Life Adjustment after the War

Though most of the "progressive" educational practices that had made their way into the schools since the 1930s were denounced by right-wing critics, more often the specific target of attack was life-adjustment education, a program of more recent origin. Formally articulated in 1945 and promoted by two national commissions in 1947 and 1954, the life-adjustment curriculum was a synthesis of many of the child-centered ideas that served as the intellectual foundation for professional educators and for the new psychological theories of social adjustment deployed by psychological experts during the war. Its goal was to provide classroom experiences that would meet the daily personal and social needs of all students, particularly those not served by the existing academic and vocational curricula at the secondary level.[31]

The perceived need for a more practical curriculum stemmed from the shifting demographics of the student population. Public school enrollments had grown continuously over the course of the twentieth century, and an increasing proportion of these students was attending high school. This was especially true during the Depression, when economic conditions made steady employment scarce. As attendance through the twelfth grade became the norm, educational leaders increasingly viewed the traditional academic curriculum as inappropriate for this new group of students.[32] Incremental changes began in the early part of the century and, over time, a more functional curriculum displaced the more academically intense college preparatory program in schools. When war enlistment and the lure of jobs in industry later produced a precipitous decline in enrollments in the 1940s, educators, seeing their ideal of universal secondary education undermined, sought to reorder the curriculum even further to increase the holding power of the high schools.[33] Developing a more personally relevant educational program for secondary students appeared to be the best way to accomplish that task.

What emerged from the various life-adjustment commissions was an outline and action plan for implementing the new program. Education professionals identified several key areas of life that they felt the successful curriculum should address, which included "citizenship, home life, physical and mental health, vocational activities, enjoyment of life, and the development of personal powers."[34] Within each area, real-life problems—problems that would naturally generate student interest—were to be the focus of instruction. Suggestions for these included everything from "getting along well with other boys and girls" to "understanding parents, driving a motor car, [and] using the English language."[35] The intent was to develop a purely functional curriculum that would apply universally to the secondary school student population.

The life-adjustment curriculum was essentially an extension of the vocational education model to general education. By making preparation for life rather than preparation for work (or college, which some viewed as preparation for professional work) the objective of schooling, the proponents of the new curricular program seized the common ground shared by all secondary students despite the diversity of their ultimate vocational goals and, in doing so, sought to extend their authority over the entire school program. Curriculum experts, through what they perceived to be a detailed scientific analysis of students and society, developed a comprehensive picture of all the tasks necessary for "the preparation of youth for the job of living."[36] With its focus on training students for this "job of living," the life-adjustment curriculum was little more than a variant of earlier social efficiency curricula prominent in the early 1900s, which were similarly designed to fit individuals to the emerging social order of the time.[37] In this mid-twentieth-century version, the emphasis was on training students to become future consumers, family members, and community participants—roles that all students, college bound or shop bound, would eventually play in the expanding consumer economy of the 1950s.

For parents and business and educational leaders looking for schools to provide greater opportunity to students in the postwar years, the life-adjustment curriculum seemed an enlightened means to educate students of all abilities and social standing to work cooperatively for the advancement of society. Its inclusive conception of education resonated strongly with democratic ideals of equality and universal participation, ideas ascendant in the 1930s and reinforced by the collective efforts of World War II.[38] Opportunities to invoke these ideals were rarely overlooked by the curriculum's proponents, who sought to match their arguments for the life-adjustment program with the strong democratic rhetoric that characterized the postwar debates contrasting the political and economic systems of the United States with those of the Soviet Union.

For these educators, to be democratic meant to be anti-elitist, and thus they worked to distance the curriculum from the historically well-established academic disciplines that were associated with the college-preparatory track. These disciplines were entrenched in the secondary-school program despite earlier efforts of educators to free themselves from their influence. Colorado education professor Harl Douglass, an apostle of the new approach, predicted that "education for Life Adjustment will mean a greatly increased freedom from subject-matter boundaries and from subject-matter organization." He explained that school will instead "be organized around problems and topics and needs, and not around the logical structure of the subject."[39] In one Montana school district the traditional academic curriculum was so marginalized that school administrators there viewed it as a program for deviants. "The scholarship plan," as it was called in the student handbook, "is better named a plan for recalcitrants." "If students and their parents believe that the only value of the high school is in learning subject matter," the handbook stated, "then the school will insist that they do more than a minimum standard. Our counseling is definitely pointed to the Life Adjustment method of graduation for all students."[40]

Although put forth as a new educational program to meet societal needs, the life-adjustment curriculum was, as already noted, not much different from earlier functional curricula. All steered clear of academic subject matter. Even when academic subjects were clearly demarcated in the schools, what was actually taught at times ranged far from traditionally-defined standards of academic rigor. In prewar high school physics courses, for example, textbooks often focused on the physics of automobiles, airplanes, and a variety of other modern machines in an attempt to generate student interest and stem already declining enrollments in the subject.[41] One 1940 biology textbook strained to find a human use for nearly every organism or, reciprocally, to provide an everyday analogy for all biological structures. Thus the student learned about yeasts because "they enable you to have tasty, light bread." Anthropomorphism was common as well; parasites in the text were described as "thieves," while a dandelion's bright flower "'advertises' for insect visitors."[42] The approach favored by publishers was to supplement a broad survey of content knowledge with an overwhelming number of tie-ins to the experiences of everyday life. The resulting textbooks, critics charged, were encyclopedic and lacked coherence; they provided teachers with too much material to cover realistically and few options for extracting a unified core of disciplinary content. Prewar textbooks of this sort required only minor modifications to meet the life-adjustment standards that were then coming into vogue.

Just as subject incoherence did not spring de novo with the life-adjustment movement, the curricular emphasis on real-life problems had its own historically rich genealogy. The life-adjustment curriculum, with its heavy emphasis on social efficiency, also possessed elements of the child-centered approach drawn from the earlier writings of John Dewey on the importance of student interest in learning. One educator who favored the new curricular philosophy observed that "as John Dewey has repeatedly pointed out, interest begets effort and a favorable mind set for learning. Therefore, the revised school program will also be more effective."[43] Dewey's ideas about education were often marshaled in support of various educational practices. They were, however, rarely implemented as originally intended. Student interest was indeed a key component of Dewey's philosophy of education, but it was only one of many conditions he thought necessary for true learning to occur.

For Dewey, education was largely a continuing process of inquiry, which originated with real problems of interest to the student that in their solution generated new knowledge useful for guiding further inquiry. The place of subject matter in this process was far from negligible.[44] But, in attempting to redress what he saw as a pernicious overemphasis on subject matter in schools—as knowledge complete unto itself divorced from social action—Dewey perhaps too zealously championed the importance of the immediate interests and problems of the student; or perhaps student interest was simply the easiest part of a complex philosophy to implement in the classroom. In any event, curricular emphasis on student interests and problems gained the upper hand in the professional pronouncements of educators during the early decades of the twentieth century.[45] William Heard Kilpatrick, an education professor at Teachers College in New York who was later to be targeted by reactionary groups in the late 1940s and 1950s for his perceived collectivist teachings, offered a new definition of "subject matter" in an article in 1923. In this new view, subject matter properly conceived was identical to life experience, nothing more.[46] Thus, life needs were converted from a way to engage students initially in the process of mastering disciplinary subject matter to the subject matter itself. The role of disciplined inquiry in the curriculum rapidly diminished, as did Dewey's enthusiasm for the direction education was heading.[47] Over time, curriculum experts arrogated to themselves the task of determining in advance the important life experiences to which students were to be exposed, relying on methods and techniques in the expanding social sciences, such as activity analysis and the school survey, to ensure the validity of their conclusions.[48]

The refinement and application of new psychological techniques during World War II helped to legitimate and propagate a new vocabulary and view of the human mind that soon added to the social science foundation of the education profession. The war significantly altered the practice of science in nearly all fields along with its relationship to society as the country pressed its intellectual resources into service to defeat the enemy. Psychologists were central to the propaganda offensives that were common after the outbreak of hostilities. In the clinical field, they developed new conceptions of mental health to aid them in their management and treatment of combat soldiers traumatized by their war experience. A key shift in strategies of psychological intervention was to focus more on the maintenance of mental health in hopes of decreasing the numbers of advanced cases of mental illness, which required more resources to treat. To this end, films, pamphlets, and books were used to counsel new recruits on the feelings and anxieties they could expect to experience in the service and to reassure them that such reactions to combat conditions were a normal and healthy sign. These materials often contained tips to help soldiers manage their anxieties—"How to Fight Fear," for example—and encouraged them to develop healthy social relations, which the experts saw as central to good mental health.[49]

The immensity of the task required psychologists to move away from highly individualistic conceptions of mental health to a mass conception that possessed a strong social component. The view held by wartime psychologists was that a healthy mental state existed when the various conflicts between the environment and personality were effectively managed through adjustment. Difficulties arose when individuals were no longer able to make such internal adjustments by themselves.[50] Though the term "adjustment" had a longer history than this, it became a popular catch-all in the 1940s, despite some misgivings about its vagueness.[51] Harvard psychologist Edwin Boring believed that the techniques of psychological adjustment developed during the war could profitably be extended to help all citizens achieve contentment in the postwar period. Self-help pamphlets like "*Fear in Battle*," he noted, "would be transformed into *How to Be Happy*," and sold at drugstores everywhere.[52] Such was the role of psychological experts, who viewed themselves as "indispensable guides in an era of social and emotional reconstruction."[53]

Goals of psychological adjustment and mental hygiene made noticeable inroads into postwar education and were central to the life-adjustment program. The real-life problems used to generate student interest in the curriculum characteristic of earlier progressive education shifted from external, more objective problems to those that were more personal. The new curriculum placed greater emphasis on the development of "attitudes, ideals, interests, habits," and "social, mental, physical, and emotional" understanding that would "enable all young people to make the most satisfying and most effective adjustment in all areas of life."[54] Educational experts based their evaluations of instructional effectiveness more on how well instruction influenced student behavior, rather than on external measures of student content knowledge.[55] New student problems included things such as, "how to succeed without bragging and to fail without making excuses" and "how to control and enjoy one's emotions."[56]

In concert with this movement to the new curricular philosophy, school districts expanded the psychological services they provided in the schools after the war in the form of guidance counseling.[57] The "keystone of the school program," one report declared, "is guidance."[58] And the proper guidance of youth was to be all encompassing, beginning in kindergarten and extending all the way beyond the secondary level, covering everything from career advising to personality development.[59] The guidance profession had, in turn, hitched its wagon to the life-adjustment program, accepting the curricular ideology therein. One guidance professional commented that "the fact that Life Adjustment Education for youth has no sharp curricular focus is of little concern."[60] The social sciences, which many intellectuals believed had fallen dangerously behind the physical sciences in the years before and during the war, had by 1949 established themselves as the foundation of the life-adjustment curriculum and the professional education community.[61] The political stability of that foundation (particularly the social psychology component), however, remained open to question.

## Concerns within the Academy

In the midst of all the political attacks being directed at the public schools in the early 1950s, the academic traditionalists headed by Illinois professor Arthur Bestor let loose with their own denunciation of the current educational trends. The postwar boom, and the infrastructure crisis it had spawned, had directed the public's attention to the material and personnel needs of the schools. The Red Scare had effectively refocused some of that attention on the intellectual substance of the curriculum, inviting debate over its adequacy. In the turmoil of the time, the academicians, perhaps sensing a window of opportunity, developed the most sustained and searching criticism of the public schools of any that decade, as well as the most telling with respect to the eventual reform of science education in the United States.[62]

The traditionalist critique was launched with the publication of the book *And Madly Teach* in 1949 by Mortimer Smith, a former school-board member and professed layman. This was followed by a tendentious address by University of Illinois botanist Harry Fuller before the local chapter of Phi Beta Kappa and the first in a long line of public education criticism from Arthur Bestor. Bestor's initial broadside came in a 1952 article in the *Bulletin of the American Association of University Professors,* the ideas of which were later incorporated into his classic polemic *Educational Wastelands.*[63] These individuals represented the interests of the new humanism in public education and made up the core of what became the Council for Basic Education in 1956. They were concerned foremost with the practices of professional educators and the impact these had on the traditional academic disciplines. Their sharpest criticisms were reserved for the educators themselves, members of state departments of public instruction and professors housed in university schools of education, whom they saw as controlling and authoritarian, pushing always to expand their methodologically suspect educational philosophy into areas previously the domain of the disciplinary experts.[64] Although public education was reeling from charges of subversion, Bestor and his colleagues focused their attacks on the inherent anti-intellectualism of the modern school curriculum—reserving particular contempt for life-adjustment education.[65] This public debate between the traditional academics and the professional educators over the proper aims and purposes of schooling brought to the fore the importance of intellectual training in society and, at the same time, highlighted the status anxiety of intellectuals in the United States. The debate also served to bring into bold relief the strong ideological climate that had come to grip the nation in all matters related to the mind.

Many intellectuals believed that the rapid growth of the social sciences in the first half of the twentieth century had come at the expense of the liberal-arts approach to understanding the human condition. It was this perceived erosion of the place of the traditional academics in American society, especially in the colleges and universities, that appears to have motivated much of the new critique of the widespread educational practices of the day. The pressing international tensions helped shape its rhetorical form.

As a general class, intellectuals historically had been marginalized in the United States. The stereotype of America as the land of inventive genius, technical mastery, and practicality worked against the establishment of any lofty place for individuals who traded in the most abstract knowledge—save when that knowledge had some direct application for the betterment of society or for personal profit.[66] This was rarely the case for those engaged in the pursuit of what are considered the more academic disciplines of the liberal arts curriculum, primarily history, philosophy, English, and foreign languages, and also the less-applied, more descriptive sciences like botany, paleontology, and zoology. What little status such intellectuals possessed reached a low ebb in the late forties and early fifties. The multiplying elements of mass culture in the postwar period not only threatened to swamp the more refined intellectual and cultural pursuits, but also in some cases even targeted and denigrated individuals of the professional and intellectual classes. The resounding defeat of Adlai Stevenson—championed by "eggheads," as Republicans contemptuously referred to them—by Dwight Eisenhower in the 1952 presidential election further reinforced the perception among academics that they were indeed under siege in America.[67]

Within American colleges and universities, the encroachment of the social sciences, which, as its proponents argued, provided a more fruitful approach to understanding and managing the human condition, met resistance in the middle decades of the twentieth century. Robert Hutchins, the president of the University of Chicago, harshly criticized what he called the "cult of science," with its "blind and unthinking empiricism," its value-free orientation and cultural relativism. He and other intellectuals called for a new humanism grounded in the classic works of Western civilization.[68]

Although providing a rational argument for a renewed emphasis on the liberal arts in American higher education, Hutchins failed to turn back the tide of social science.[69] As already noted, the expansion of social-science techniques, indeed, accelerated with the onset of World War II. In the context of world war, a reasoned understanding of the human condition, not surprisingly, was ignored in favor of the ability to manage massive amounts of information regarding the social and psychological characteristics of American manpower resources. Instrumental needs such as these required the empirical techniques science offered.

Concerns over the validity of the social sciences, as applied to the daily affairs of people, were echoed in the traditionalist denunciations of the life-adjustment curriculum. Mortimer Smith in his book, *And Madly Teach,* declared that "too much of the science of the mind has become bad science. . . . It would be unfair to condemn modern psychology in toto, but certainly the *general tendency* . . . is toward dogma; it is not that it is getting away from its prime function, which is study and investigation of the mind in order to discover general principles; but it is adding another function, namely, that of deducing pseudoscientific data from those principles that, if we will only follow them, will reward us with happiness, success, and satisfactory sex lives."[70]

The threat to the established academic curriculum by the social science-wielding professional educators appears to have been personally felt by a number of university scholars in the early 1950s, particularly in the form of education school requirements that served to bypass courses in the liberal arts. They argued that the need to reestablish academic standards was demonstrated by the grossly inadequate intellectual preparation of first-year college students.[71] The desire of the education establishment to spread the ill-conceived, life-adjustment practices that had devastated secondary education to the university level promised only to exacerbate the situation. This encroachment, more than anything, likely prompted their vigorous defense of the academic curriculum. Educators in the early part of the century had gained control over schooling by convincing the state of their expertise in educational matters.[72] Newly formed teachers colleges trained educators, who, when properly credentialed, were employed by local and state governments in various departments of education—departments that, in turn, set public school guidelines and established certification requirements that prescribed the course of study for future teachers. These educators in state bureaucracies, universities, and public schools constituted what Bestor referred to as the "interlocking directorate"—concerned primarily, he argued, with the self-interested perpetuation of their own authority.[73]

This "directorate" had little direct control over the content of the local school curriculum. Curricular decisions were made by individual teachers and school boards, perhaps influenced by but rarely consisting of education professionals. Their real power lay in controlling the state-sanctioned certification process, which, according to Bestor, deliberately sought to eliminate traditional academic coursework from the teacher education programs in favor of vacuous work in pedagogy, or methods. The beneficiaries of this, he explained, "are the professors of education, who are thus assured of a steady flow of students through their courses."[74] He feared not only the effects of this trend on the quality of teacher education, but also the further marginalization of the academic subject areas. Educators, Bestor charged, "look forward to the ultimate subversion of science and learning even in colleges and universities. A few have gone so far as to draw up blue-prints for conquest, describing the future institution of higher education after the educator shall have made it over in his own image."[75]

### Bestor's Critique

To check the spread of this anti-intellectualism and ultimately safeguard the place of the liberal arts in higher education required a well-thought-out offensive on the part of Bestor and his like-minded colleagues. Developing a credible attack on public education, after all, was dicey business, especially in the early 1950s. The public school was the prototypical American institution, and with the dramatic increases in student enrollment came a vocal parental constituency that, for the most part, was satisfied with the schools their children attended.[76] Many parents, characteristically distrustful of intellectual elitism, favored the functional curriculum that schools adopted as it was. Furthermore, the NEA sought to portray those critical of public schools as unstable reactionaries, all to be lumped together with the likes of the right-wing zealot Allen Zoll.[77] Attacks on public education were easily cast as attacks on long-held American democratic ideals.

Complicating matters was the position of weakness from which the academic traditionalists launched their attack. The status of intellectuals in the United States clearly was low. Hearings held by the Illinois School Problems Commission in 1952, before which Bestor and other faculty testified concerning the dangers of the professional education monopoly, amply bore this out. At one point in the hearing following Bestor's testimony, a state official, in reference to an unrelated matter, derisively inquired whether it had been "approved by those men in liberal arts." The next day's headline read, "Board Jokes Over Battle of Educators."[78] Disrespect easily begat suspicion. Following the startling conviction of Alger

Hiss and the repeated smears of intellectuals as communist sympathizers by the radical right, members of the academic community had the added obstacle of maintaining an acceptable ideological stance in controversial matters—a point of which Bestor was well aware.[79]

Bestor's first attempt to take action came in a proposal submitted to the Council of the American Historical Association in December 1952 that called for the establishment of a scientific and scholarly commission to help ensure the adequate representation of academic subjects in the school curriculum.[80] Colleagues within his own discipline, fearful of the threat red hunters like Senator McCarthy posed to academic freedom and their own careers, urged Bestor to proceed cautiously. There seemed to them too much criticism of the schools already. In response to this proposal, Harvard historian Arthur Schlesinger expressed just that sentiment. Bestor, however, remained steadfast. "I agree that McCarthyism applied to education is even more dangerous than this," he replied to Schlesinger. "If my resolutions seem to show no concern about the matter, it is simply because I did not want to fight two battles in one set of resolutions. I think we must fight the menace in two ways: by direct resistance, and by preventing the undermining of respect for free intellectual inquiry."[81]

For Bestor to make a compelling case for a return to the traditional academic curriculum in the social and political climate of the 1950s, it was clear that he had to thread his argument carefully between images and ideas that would likely draw charges of either communist sympathizing or antidemocratic elitism.[82] Not surprisingly, with Americanism the order of the day, Bestor heavily draped his writing with the Cold War rhetoric of democracy—the one societal ideal, though the foundations of which were often debated, above reproach in the postwar United States.

In *Educational Wastelands,* Bestor's most complete and well-publicized diatribe regarding the inadequacies of public education, he turned the charges that the academic curriculum was antidemocratic back on the professional educators, working to place them outside the current ideological mainstream. Bestor was strongly committed to the belief that schools were institutions primarily responsible for the training of the intellect, and that the discipline-centered curriculum was the most effective means to accomplish that.[83] He was especially disturbed by the fundamental assumption inherent in life-adjustment education that intellectual training, as traditionally conceived, was inappropriate for all but a small minority of the population. On the contrary, Bestor argued, "Popular education is designed to endow the people as a whole with precisely the kinds of intellectual power that have hitherto been monopolized by an aristocratic few." A curriculum that provided such training was not only democratic in that it provided every citizen with "knowledge, cultural appreciation, and disciplined intellectual power," but also absolutely necessary in a democracy, where in self-government the "functions of an aristocracy become the functions of citizens at large."[84] The functional curriculum pushed by the educational establishment in fitting students for their projected roles in life, Bestor argued, sold short their potential for intellectual development and greater democratic participation. The result would be greater dependency on elite groups of professional experts. He explained that "once the fancy rhetoric is stripped away the argument for lax academic standards always turns out to involve plain, old, condescending, anti-democratic distrust of the common man and his intellectual capacity."[85]

Cultural enrichment and its concomitant increased respect for the intellect was a key part of Bestor's vision. "If the schools are doing their job," he maintained, "we should expect . . . a significant and indisputable achievement in raising the intellectual level of the nation—measured perhaps by larger per capita circulation of books and serious magazines, by definitely improved taste in movies and radio programs, by higher standards of political debate, [and] by increased respect for freedom of speech and thought."[86] The alternative that Bestor and other academics offered was firmly grounded in the liberal arts tradition. Subjects such as history, literature, foreign languages, mathematics, and the sciences were valuable and should be taught, they argued, not for any immediate functional reasons in the most practical sense, but rather for their ability to communicate to students the "cultural values of a nation."[87] The promulgation of these values would later become one of the primary objectives of the curriculum projects supported by the National Science Foundation.

Bestor's colleague, the botanist Harry Fuller, lamented the elevation of the functional over the humanistic emphasis in the sciences in particular. Voicing a complaint that would be heard throughout the 1950s and 1960s, he declared: "Biology courses may still bear the tag of biology, but their content is often reduced to personal hygiene, what to do about forest fires, how to breed better corn and sheep." He went on, "Gone from many such biology courses are the basic, the impressive, the truly significant biological phenomena: the panorama of life through geologic time, the marvelous interrelations of tissue structure and function in living bodies, the wondrous adaptations of flowers to pollinating insects, the mysterious migrations of birds"—all subjects of scientific study that, though not meeting any trivial personal needs, certainly contributed to a greater understanding of the place of humanity in the grand picture of life on earth.[88]

More important for Bestor's argument than the cultural enrichment a traditional academic curriculum would provide, however, was the role of disciplined inquiry, long lost from the curricular prescriptions of the professional educators. In his view, the disciplines "must be presented . . . as systematic ways of thinking, each with an organized structure and methodology of its own." Such thinking certainly did not consist of the mere memorization of subject matter, the placing of facts into the "cold storage" of the mind for later use. This metaphor, repeatedly invoked by critics of the liberal arts, was particularly disagreeable to Bestor. "The liberal disciplines are not chunks of frozen fact. They are not facts at all," he explained. The traditional academic disciplines, he went on, were "the most effective methods which men have been able to devise, through millennia of sustained effort, for liberating and then organizing the powers of the human mind."[89] Each discipline is unique in the active methods of inquiry it employs in making sense of the world. Thus, problems of quantity and relationships are dealt with using mathematical intellectual tools, those of the past with historical techniques and reasoning patterns, and understanding matter requires the scientific inquiry inherent in the practice of physics and chemistry. According to Bestor, mastering these various methods of inquiry entailed learning the "inner structure and logic" of the disciplines.[90] Since the purpose of schooling, for Bestor, was to cultivate the power of disciplined thought, the curriculum must perforce consist of the traditional academic disciplines that are alone capable of conferring that power.

The value of such thought went well beyond what it might contribute to cultural elevation or enlightened democratic participation. As important, perhaps, was its capacity to combat communism in the ideological contest then being waged. The life-adjustment curriculum, its critics insisted, was dangerous in its dependence on a sociological theory of education that had the potential, as Smith charged, to "mold youth in any desired shape, toward any ideology" using techniques of mass manipulation common to totalitarian societies. In contrast, the humanist alternative with its emphasis on methods of rigorous inquiry had the power, Bestor claimed, to liberate human minds.[91] In discussing the treatment of controversial subjects in schools, by which he meant those that wandered close to areas some thought subversive, Bestor employed the same reasoning as Sidney Hook: that objective inquiry was incompatible with communist ideology. He explained that "to train a man to think clearly and originally is nothing else than to train him to handle controversial issues," and that the appropriate engagement with such issues simply requires "the methods of free, objective, critical, scholarly inquiry. No other methods are legitimate and no others can safely be tolerated in an educational system founded on the principles of responsibility and freedom." Continuing to echo Hook, he declared, "We have no obligation— indeed, no right—to tolerate the intellectual methods of the communist who has abandoned free, critical investigation out of devotion to a party line."[92] Thus Bestor cast the traditional academic curriculum, democratically tailored for all students, as ideologically pure and life-adjustment education as ideologically suspect.

In laying out his argument, Bestor, often the target himself of extremist invective, was not one to shy away from what might be considered inflammatory, if politically expedient, rhetoric. Though maintaining his focus on the anti-intellectualism of the modern curriculum, he did not hesitate to draw attention to the subversive characterizations already made. He described the primary technique employed by professional educators as nothing more than indoctrination, and, invoking a particularly disturbing image, explained that such techniques serve only as "a narcotic to kill the pain of thinking. And it prepares a student to face new problems in only one way—by reaching for the hypodermic needle." His section on the educational establishment was firmly situated in the political paranoia of the time. "Across the educational world today," he wrote, "stretches an iron curtain which the professional educationists are busily fashioning. Behind it, in slave labor camps, are the classroom teachers, whose only hope of rescue is from without. On the hither side lies the free world of science and learning, menaced but not yet conquered."[93] The provocative language Bestor used was no doubt a testament to his convictions regarding the fundamental importance of reestablishing high standards of intellectual training in American public schools. It also reveals the pervasiveness of the images and vocabulary of Cold War conflict in the public consciousness, a conflict that rapidly intensified in the period from the late 1940s into the 1950s.

For all its initial appeal, the life-adjustment education program, whether the result of the growing criticism or its own excesses, eventually fell from public favor. Its emphasis on teaching for the trivialities of daily life had reached extremes that few were willing to publicly defend. One senior high school psychology textbook, for example, included "how to 'take a joke'" among the "important things which each individual has to learn" in school. To master this social skill, the author explained in all seriousness, "takes some training, effort, and experience."[94] Writing in a professional journal, another educator baldly declared that "We shall some day accept the thought that it is just as illogical to assume that every boy must be able to read as it is that each one must be able to perform on a violin, that it is no more reasonable to require that each girl shall spell well than it is that each one shall bake a good cherry pie."[95] As education historian Herbert Kliebard has observed, "an expert marksman like Bestor had little trouble hitting such a grossly inflated target."[96]

More significant in the downfall of life-adjustment education, however, was the profound shift in the ideological climate in the postwar United States. Rising national paranoia threw cold water on any public policies that smacked of collectivism. In this light, critics viewed the life-adjustment program as merely an extension of the already suspect progressive education practices.[97] In addition, the program's emphasis on adjusting the individual to society reflected its reliance on an intrusive social science that had overreached its base and brought to mind the manipulative techniques of mass persuasion and brainwashing that were closely identified with the totalitarian nations in the enemy camp. The emerging spirit of the country, in harmony with the conservative political resurgence, was one of laissez-faire individualism. America in the 1950s, as historian James Patterson describes, was brimming with opportunity: "it was defined by the belief that hard work would enable a person to rise in society and that children would do better in life than parents."[98] With the seemingly unbounded prosperity, concern for the collective welfare gave way to faith in a new meritocracy where individual excellence was increasingly identified with national strength.[99] The intensification of the Cold War, with its greater reliance on sophisticated technology as a means of ensuring national security, placed a newfound premium on intelligence, if only of a technical sort. Life-adjustment education was hard pressed to find any purchase in this new postwar world.

Despite Bestor's impassioned calls for a renewed commitment on the part of scientists and scholars to reinvigorate the secondary school curriculum and the general public outcry for federal aid to education, by 1954 things remained relatively unchanged in schools across the nation. Although life-adjustment education was no longer a real force, the promise of a new curriculum along the lines the university scholars such as Bestor and Fuller suggested remained unfulfilled. The decentralized nature of the United States school system made systematic reform nearly impossible, and the strong tradition of local control effectively blocked any federal initiatives in this area. What was left were the persistent ideological and military threats of international communism and American public schools revealed as inadequate to the task of preparing the next generation to meet these threats. One might say that ground had been broken, but a new educational edifice had yet to be built.

# CHAPTER 2

# THE STATE OF SCIENCE IN AMERICA

When Arthur Bestor called for the establishment of a national commission of "scientists and scholars" in 1952 to evaluate school curricula and, presumably, restore the subject-matter disciplines to their rightful place, little note was taken of his explicit and repeated inclusion of scientists as key participants in the process.[1] It appears somewhat odd that a proposal designed to safeguard the professional status of scholars in the humanities, drafted by a historian, and presented at the annual meeting of the American Historical Association, would give scientists top billing. Considered in light of the relative prestige accorded the two professions, however, Bestor's actions make more sense. After World War II, scientists had garnered for themselves a considerable amount of positive public notoriety. With intellectuals suffering under the withering ideological scrutiny of McCarthy and other right-wing zealots, Bestor sought to forge a key alliance with the much more reputable scientific community as a means of lending greater legitimacy to his proposed educational reforms.

Bestor and other academics mulled over this strategy at the Barclay Hotel in New York City at the planning conference for the Council for Basic Education. During this meeting, Bestor stressed the importance of securing business-community support for their project but worried about how that might be accomplished, fearing that businessmen would be reluctant to associate themselves with potentially subversive academics. But, "if we could get the scientific societies and those of the humanities together, to create an organization which would be above suspicion," he proposed, "it would be in a way immune to the kinds of smear tactics which could be launched." Harry Fuller, the Illinois botanist, concurred, explaining that such an approach would draw attention away from the perceived "queerness" of the intellectuals. "It gives us strength and support," he added, "if we can get other groups, who are in American culture regarded with perhaps more respect" and "are taken more seriously."[2] Both Fuller and Bestor early on had cast about for such support in the scientific community, publishing their educational critiques in *Scientific Monthly* magazine, one of the official publications of the American Association for the Advancement of Science (AAAS).[3]

By the 1950s, scientists did indeed enjoy high prestige in rather tense ideological times. And, scientists' complaints notwithstanding, the public responded with a general gravitation toward scientific topics and issues, leading one academic to wonder aloud why it was that "the scientific work in schools is so much more satisfactory than the humanistic." After all, he declared, "there is nothing inherently more fascinating about protoplasm or atomic energy than there is about John Donne or the Taiping Rebellion."[4] A great deal of this attraction can be attributed to the technological wonders science had produced. But part of the appeal, perhaps, also lay in the perceived uncontroversial nature of science itself. Here was an area of study that could claim access to objective truth about the natural world during a time when the truth about many things was ambiguous at best. Science was a subject untainted by the distortions and subversions to which the humanistic studies were susceptible. In a world of competing ideologies, science was seen by many as simply nonideological.

At a surface level, science has always been identified in the public mind with objectivity, the disinterested pursuit of truth. Probing more deeply, however, reveals a much more complex picture regarding the public perception of science in the years after the war. It was a picture that laid the groundwork for the scientists' involvement in education reform well before the Soviet launch of *Sputnik,* the event commonly thought to have prompted their efforts. Some saw it as an objective method of inquiry, technological mastery of nature, the key to Allied victory—all of which, to some extent, had positive connotations. Science was at times also associated with the social and political activities of various scientists and new methods of social engineering, which, in the political climate of the time, were viewed with suspicion. As scientists were drawn into more intimate working relationships with the federal government, these perceptions held by the general public and, by extension, Congress and other government officials increasingly had a direct impact on the conditions under which scientists worked.[5]

The conflation in the public mind of science and technology was the most vexing of the challenges facing the scientific community. The new instrumental technologies scientists developed during the war brought them generous government patronage and national status, which they welcomed.

But the success of these technologies also worked against their efforts in the postwar period to maintain public funding for research that seemed at times far removed from the country's day-to-day economic and social needs. This confusion contributed to a social and political environment that scientists believed more profoundly threatened the autonomy of their work. As respected as science may have been, scientists, like other intellectuals, were targeted by reactionaries both in and out of government during the turbulent years of the Red Scare. With their technological contributions to national defense, they were, in fact, subject to even greater ideological scrutiny than most. As members of an elite group in a country with a strong populist sensibility, they had to work hard to overcome public resistance to the privileged position science and scientists seemed poised to claim in the new era. Their efforts centered on carefully distinguishing basic research from technological development and cultivating a perception of American science as inherently democratic in its organizational structure and virtuous in its pursuit of truth. These efforts to reshape the public image of science, whether ultimately successful or not, illustrate the key themes that later became central to the view of science portrayed in the PSSC and BSCS curriculum materials—a politically palatable image of science that the federal government and the public at large could comfortably support.

## Science and the War

World War II was a watershed for a variety of groups and institutions in the United States; for the scientists, it served as their proving ground. Organized by MIT vice president Vannevar Bush through the Office of Scientific Research and Development (OSRD) and sustained by nearly unlimited federal funding, the scientific community developed new military technologies that contributed substantially to the Allied victory. Military leaders, though initially indifferent to the role of scientists in the war, quickly recognized the advantages of harnessing their powers to meet their needs. The scientific research community, by demonstrating their technological proficiency, thus guaranteed themselves a ticket to the inner workings of the government's war effort and set the stage for both their postwar public reception and their subsequent role in maintaining the security of the United States.[6]

Scientific contributions to the war effort included things such as the proximity fuze developed at Johns Hopkins, solid-fuel rockets engineered at the California Institute of Technology, synthetic rubber, and the wide-scale production of penicillin. The most significant of these wartime scientific triumphs in the eyes of many scientists and military officials alike, however, was radar. Developed in consort with the British at MIT's Radiation Laboratory (deliberately named to suggest militarily inconsequential, or so it was thought at the time, research in the field of nuclear physics), radar systems utilized high-frequency electromagnetic waves in the centimeter range to locate distant objects. In their various forms, these systems were capable of the seemingly impossible task of seeing far into absolute darkness, dense fog, or smoke, making the invisible visible. Such sensory enhancement proved invaluable to those on the front lines in warning of approaching enemy planes, locating German submarines prowling the waters of the North Atlantic, directing anti-aircraft fire toward their targets, guiding large-scale bombing missions, and ensuring the safe landing of aircraft in poor weather conditions.[7]

Of course, few technological developments, however crucial they may have been to the Allied effort, could compare in raw power or imagery to the atomic bomb. Under the direction of the theoretical physicist J. Robert Oppenheimer at Los Alamos, Manhattan Project scientists had succeeded in producing a nuclear-fission weapon that delivered the explosive power of over twenty-thousand tons of TNT in a single, blue-white fireball several hundred feet in diameter and several million degrees in temperature. The detonations of these weapons at Hiroshima and Nagasaki obliterated nearly every structure in the immediate vicinity of the blasts. When the characteristic mushroom cloud of dust and smoke had cleared over the city of Hiroshima, four and a half square miles of it had been leveled. It looked, one observer noted, as though it had been "ground into dust by a giant foot."[8] Together the bombs killed over one hundred thousand people, injured thousands more, and brought the war to a swift conclusion. Postbombing surveys indicated that the destruction inflicted by the new weapons in those two explosions would have required 330 B-29s each carrying ten tons worth of conventional bombs.[9]

With the shroud of secrecy lifted at war's end, the awe-inspiring, almost magical, accomplishments of the scientists were widely hailed in the popular press. Such accolades led to unprecedented levels of public status. "The near veneration of atomic scientists in the later 1940s," historian Paul Boyer has noted, "is one of the most striking features of this period."[10] Even as the public recognized their contribution to a new form of global terror in the bomb, many derived some comfort from the fact that the scientists responsible for unleashing nature's power did so for the cause of democracy and the people of the United States. Indeed, the image of scientist as savior had a strong hold on the public consciousness. *Life* magazine devoted nearly an entire issue to science in the war, during which "the 'absent-minded' professors with their theories of relativity and interminable formulae shed their black alpaca coats and overnight . . . donned the tunic of superman."[11]

In the ensuing years, the estimation of scientific expertise more generally seemed to outstrip any realistic appraisal of its value. A new wondrous age seemed just around the corner—indeed many described the postwar decades as the dawning of a new "Age of Science"—one that promised freedom from diseases such as cancer, unlimited supplies of power, and technologies that would provide relief from most every human problem.[12] Physicists in particular, befitting their role in the war, commanded the bulk of public attention. It appeared that "no public forum on the issues of the nuclear age [was] complete without a physicist," wrote historian Daniel Kevles. "Physicists were asked to address women's clubs, lionized at Washington parties, and paid respectful attention by conventions of theologians and social scientists."[13] They were encouraged to profess outside their narrow areas of expertise, and often did so despite warnings handed down by some senior members of the scientific establishment.[14]

The obvious military utility of science during armed conflict, as troubling as its products might have been, contributed to a faith that technological mastery could safeguard America through the Cold War as well. At the turn of the decade, fear of communist expansion by the Soviet Union far outpaced fear of nuclear annihilation, prompting public demands not that nuclear technology be scaled back, but rather that nuclear and general technological superiority be pursued with all deliberate speed. As Boyer noted, "Americans now seemed not only ready to accept the bomb, but to support any measures necessary to maintain atomic supremacy."[15] The political decision by the Truman administration to combat communist ambitions with massive remilitarization placed an ever greater premium on the technical expertise of the scientific community. Many scientists were more than happy to provide their services in the cause of national defense.[16]

In stark contrast to the paucity of resources being made available to the nation's schools, the federal government was directing an enormous research enterprise by the beginning of 1950.[17] The Truman administration had committed the United States to a defense strategy built upon a preponderance of sophisticated, evolving military technology, which many believed was the best means of countering the Soviet manpower and conventional force advantages it then held.[18] The outbreak of the Korean War that summer greatly intensified government research efforts, and to better manage this increased reliance on the R&D establishment, the president established the Science Advisory Committee within the Office of Defense Mobilization (SAC-ODM). Scientists and research administrators such as MIT president James Killian, Robert Oppenheimer, Alan Waterman (head of the newly formed National Science Foundation), and Jerrold Zacharias thus became "key players in postwar national security policy making."[19]

The bonds between the national security establishment and the major research universities grew tight as the decade wore on. Truman's foreign policy stance was further reinforced with Eisenhower's election in 1952. One of the first national security initiatives of his administration was the unveiling of the "new look" defense program, which was designed to exploit the cost effectiveness of advanced weapon systems (thermonuclear weapons in particular) in order to produce more "bang for the buck" in hopes of alleviating some of the financial pressures the Cold War had brought to bear on the federal budget.[20] The consequence of this strategy was an even greater reliance on the scientific talent of the country. To facilitate consultation in these matters, the Eisenhower administration, under the direction of Arthur Flemming (head of the ODM), brought the Science Advisory Committee even further into the inner circles of national security decision making.[21] This "mutual embrace" of science and the military marked the origins of what historian Walter McDougall and others have called a technocratic political organization in the United States: "the institutionalization of technological change for state purposes."[22] Such purposes, as we will see, were not limited to the military.

## Public Misunderstandings

Many scientists relished their newfound power and respect in the decades following the war. They not only profited from the open spigot of federal dollars, but helped to direct the flow of funds from their positions on various government research advisory boards. "The short-term benefits of this partnership—bigger budgets, better facilities, more political clout in Washington, ever more sophisticated military hardware, and even a few Nobel prizes"—noted historian Stuart Leslie, "were obvious to everyone."[23] Research projects employed growing teams of scientists that brought complex new laboratory techniques and equipment to bear on questions about the natural world.[24] In a short few years researchers "had grown accustomed merely to signing an order for a new instrument whose purchase might have required a major faculty debate before the war."[25] University administrators had also come to depend heavily on the flow of government money as a means to supplement revenues and maintain, or even improve, their position in the national research hierarchy.[26]

Though scientists could measure their esteem in research dollars, it was an esteem limited in many ways to their narrow technological proficiencies; the luxuries they enjoyed in the laboratory came with some professional cost. As early as 1946, Cornell physicist Philip Morrison, later a key player in the PSSC curriculum project, had already sensed the price to be paid: the loss of scientific freedom of inquiry. Though many military agencies amply funded basic, or nondirected, research, the underlying motive for such generous patronage was all too clear. Morrison attempted to warn the scientific community of the impending loss of control. The "now-available contracts," he wrote, "will tighten up and the fine print will start to contain talk about results and specific weapon problems. And science itself will have been bought by war, on the installment plan."[27]

Whether the opposition to government control was due to concern over the military ends of scientific research or to a vague discomfort with government intervention in general, the scientific community found itself precariously positioned as an intellectual elite in a democratic society. Their dilemma was a powerful one: in the eyes of scientists, the advancement of scientific knowledge depended on the intellectual freedom (of scientists) to pursue whatever research seemed most promising; however, in the new postwar research economy, it also depended on substantial government funding allocated by a Congress and public concerned with practical results, especially as those results played out in the technological competition with the Soviet Union.[28] The goal of the scientific community was obvious, if not easily accomplished: to maximize funding, while minimizing federal control of their work. Managing the public perception of science increasingly became an important part of their efforts to secure public acceptance of the scientific worldview, which would bring with it, they believed, these favorable conditions for scientific research. Unfortunately social trends seemed to be working against them. Resistance to the scientific attitude, which scientists saw as a fundamental irrationalism, seemed all too prevalent among members of the general public and appeared to be on the rise.

By the early 1950s, the intensification of the Cold War and the fear of domestic subversion contributed to public anxieties that seemed to push many toward antirationalist worldviews. The movement toward religious transcendentalism (clearly in opposition to science) among conservative intellectuals spilled over into the masses. A groundswell of religiosity in the United States led to an explosion in church construction. From 1946 to 1950, funds expended for this purpose increased over 400 percent. A significant portion of this new church membership filled the ranks of the previously marginal fundamentalist and evangelical denominations. In addition to the solace many found in the certainty of sectarian dogma, the strong identification of communism with atheism made religious participation a particularly visible way of demonstrating one's patriotism. Congress, always a mirror of public sentiment, added legitimacy to the public embrace of "spiritualism" by voting to add the words "under God" to the pledge of allegiance in 1954 and that same year mandated that all U.S. currency carry the motto "In God We Trust" as a bulwark against godless communism. This religious resurgence was viewed by some as a growing threat to the advancement of scientific thinking.[29]

More direct challenges to scientific authority were evident as well. The best example is found in the controversy surrounding the systematic fluoridation of the public water supply.[30] Though the power of artificial fluoridation to prevent tooth decay was well established epidemiologically in the 1940s, local groups across the country angrily objected to government tampering with their water. Many such groups charged that fluoride, even in small amounts, was a nerve poison or a carcinogen; that fluoridation programs amounted to compulsory medication and thus violated individual rights; or even that it caused an increase in juvenile delinquency and sexual perversion. The public outcry was loud enough to prompt a congressional investigation into the safety of fluoridation programs in 1952, which did little to alleviate public concern. Continued opposition in local communities prompted the editors of *Scientific American,* concerned with the rejection of scientific authority, to lead off their 1955 issue with an article entitled, "A Study of the Anti-Scientific Attitude," which attempted to characterize the nature of these "bizarre, irrational fears" exhibited by the public.[31] The study found that suspicion of science was particularly high among those with less than a high school education, individuals who were often influential in local affairs through the sheer numerical advantage they possessed. One professor at Williams College, summarizing the fluoridation controversy in his town, concluded despondently that "irrationality in areas of morals and politics," had now spread to "areas where science usually holds sway."[32]

The rapid development and testing of nuclear weapons technology occurring at this time only added fuel to the antirationalist fire. The explosion of the United States's first hydrogen bomb on the Bikini Atoll in the Pacific in 1954 touched off a storm of controversy after it became known that the crew of a Japanese fishing boat nearby had been contaminated by radioactive fallout from the blast.[33] A series of atmospheric tests in Nevada the following year produced radioactive rain in Chicago and a radioactive cloud that stretched from Nebraska to New York. This dangerously inclement weather generated considerable public hysteria despite the fact that, according to a recent Gallup poll, only 17 percent of the public understood what fallout was. The combination of fear, hostility, and ignorance in large segments of the population was understandably disconcerting to the scientific community, which depended on popular support—or at the very least acquiescence—in order to maintain federal funding of large-scale scientific research. As one scientist openly conceded, "if the public pays the bill, then what the public thinks science is may . . . have a direct bearing on what scientists are permitted to do."[34]

These concerns over the relationship between public support and public control of science were laid bare in the debates over the establishment and operation of the National Science Foundation—the federal agency that provided the key institutional base for the science education reform movement. The Foundation had been created in 1950 as an independent funding and coordinating agency for scientific research. It had initially been conceived as an institution dedicated to harnessing science for the immediate social and economic needs of the nation following the war. Hard lobbying by scientists themselves and a good deal of congressional debate, however, resulted in a science agency largely insulated from public direction. The political organization of the new foundation conformed to the vision laid out in Vannevar Bush's 1945 report, *Science—The Endless Frontier.* In this document, Bush proposed an institutional arrangement that would provide long-term funding for scientific research and education to be allocated at the discretion of the scientific community rather than the public. As one researcher put it, what the scientist needed was "the intellectual and physical freedom to work on whatever he damn well pleases."[35] That this work was to be funded by the federal government seemed to many beside the point.[36]

The key to Bush's argument for scientific autonomy rested on the relationship between basic, or pure, research and applied research.[37] The definition he provided for basic research—research "performed without thought of practical ends" that "results in general knowledge and an understanding of nature"—though useful for insulating scientific work from political interference, at the same time promised little public payoff and, consequently, little congressional support.[38] Bush, and NSF officials frequently thereafter, explained that technological applications, such as those successfully developed during the war, depended directly on the reservoir of fundamental scientific knowledge so far accumulated.[39] The formula was clear, if somewhat ironic: funding in the least practical areas of science, the "purest realms" in the words of Bush, was the most direct path to economic and technological progress. Thus, by maintaining the primacy of basic research as "the pacemaker of technological progress," scientists were able to ensure that the practical needs of the state would not interfere with the support of scientific research.[40]

Historian Daniel Kevles judged the outcome of the NSF struggle as a "victory for elitism."[41] And, indeed, the elite members of the scientific community as a whole had succeeded in establishing their vision of the proper institutional place of science in postwar America. These individuals were primarily physicists who cut their teeth on research administration in the OSRD, at MIT's Radiation Lab, and Los Alamos. Members of this group dominated the National Science Board, the governing body of NSF. The first director of the Foundation, Alan T. Waterman, was himself a physicist, a former staff member of the OSRD, and, prior to taking over his duties as director, had been the chief scientist for the Office of Naval Research. The National Science Board wasted little time in implementing this plan. At the board's first meeting in 1951, despite a rather broad legislative mandate, board members voted to limit the Foundation's activities solely to the advancement of basic scientific research and training.[42]

With the financial support of basic research as its primary mission and with its appropriations visible to all in the public forum of congressional committees (rather than being buried in the enormous defense budget as were funds for most other scientific research projects), the Foundation's level of funding became an early barometer of public commitment to pure science. NSF officials, as well as other members of the scientific elite, were shocked by the low reading. The National Science Board knew that it had to work hard to make basic science "seem valuable to the public" in order to avoid "ever diminishing budgets."[43] The Foundation had nevertheless expected to receive its full $15 million request for fiscal year 1952, the agency's first in full operation. Instead, the House Appropriations Committee, convinced that basic science held out small promise, especially during the Korean crisis, offered only $300,000—a dramatic 98 percent reduction in funds.[44] "Scientists across the country, as well as the *New York Times* and *Washington Post*," historian James Hershberg described, "shrieked in outrage at so blatant a sign of congressional contempt."[45] Only after Alan Waterman and other scientists made a last ditch plea, heavily emphasizing the contributions basic research would make to defense-related technology, did Congress adjust upward its initial appropriation recommendation to $3.5 million.[46]

The less-than-expected funding for NSF had little impact on the overall level of scientific research activity. In the postwar period prior to the Foundation's establishment, numerous defense agencies had generously filled the funding vacuum. Of greater concern to the leaders of the scientific community, however, was the clear message sent by Congress regarding the perceived value of basic scientific research. It was a message somewhat surprisingly echoed only a year later by the Secretary of Defense Charles E. Wilson, who condescendingly observed that "basic" was what you called research "when you don't know what you're doing."[47] According to Eugene Rabinowitch, editor of the *Bulletin of the Atomic Scientists,* such disrespect for basic science clearly boiled down to a government with "not merely an indifference to science, but a sub-conscious hostility to it."[48] This same frustration with science's reluctant patron was evident in Vannevar Bush's comments to Detlev Bronk, president of the National Academy of Sciences (NAS). "What a grand thing it would be if the members of Congress understood better the relations of the Federal Government to fundamental research!" he exclaimed. "They understand these better than they did in 1940, to that I can testify. But they acquire this understanding in the same way as other citizens," through education. The nature and scope of such education was hard for Bush to conceive, however. The NAS could hardly undertake such a campaign for, in his view, "it would soon be on the griddle if it did, and there is much doubt if Congress would listen."[49]

### Security Restrictions and McCarthyism

Although scientists won the battle over control of the National Science Foundation, it was but a small victory. The continued pursuit of basic scientific research at public expense remained far from assured. Even as science continued to contribute impressively to the material welfare, health, and security of the country into the 1950s, a variety of cultural and political obstacles increasingly interposed themselves between scientists and their grand vision of government patronage. According to McDougall, there were essentially two kinds of challenges to the "reigning American culture of technology," which equally threatened the scientific research enterprise: "the first was represented by all those who ever asserted that technology or 'modernity' were offensive to other [mostly religious] values," and the other "came from would-be technocrats, Marxist or otherwise, who advocated far greater public promotion of technological change."[50] The latter challenge, from those who sought to bend science to political and social ends, had deep roots in America. In the postwar period, this challenge manifested itself in government attempts to direct scientific research to meet national needs. The former challenge, from those who sought to free modern society and culture from the tightening grip of the scientific worldview, was a relative newcomer—a product of the mid-twentieth century atrocities and anxieties.[51]

The scientists' frustration with the repeated wrangles over funding did not even begin to express their deeply-felt concern with the overall relationship between science and the state. More pressing was the persistent interference in the affairs of the scientific community tied to the rising paranoia over national security. Of all the challenges to science in the late forties and fifties, in the eyes of scientists, perhaps none surpassed the Red Scare in importance. The Red Scare, which had run rampant in education, was unique in its embodiment of both fundamental threats to science— the efforts by the government to control the professional activities of scientists and the growing public irrationalism that seemed to jeopardize the possibility of social progress driven by a scientific worldview. Reaching its peak in the early 1950s, this crusade against domestic subversion left an indelible mark on the scientific community that would shape its actions for years to come.[52]

Scientists had their first taste of security restrictions with the crash programs of World War II, but interestingly it was during the following peace when such measures were uncomfortably tightened. At Los Alamos, scientists had come to accept certain precautions necessary to conceal their work from the enemy. They tolerated the geographic isolation and the constant presence of security officers shadowing their every move. They balked, however, at the military's insistence on the intellectual compartmentalization of laboratory research, which scientists believed was an unnecessary barrier to progress.[53] The successful production of atomic bombs seemed to indicate that they knew what they were talking about. The free exchange of information that was central to the scientific ethos and tolerated at Los Alamos for the purposes of expediency was sharply curtailed, however, in the postwar period. The range of measures taken by the government to restrict the flow of information was truly remarkable. Not only were scientists given limited access to classified information related to both military and nonmilitary projects, but the normal social interactions of scientists were restricted as well. As one MIT professor complained, "security procedures have disrupted the freedom of travel and of scientific intercourse, in fields of basic science where no question of secrecy was involved. . . . [S]cientists requesting visas to give lectures or to attend professional meetings . . . have experienced frustrating and expensive delays and in some cases have been denied visas."[54]

If the general security conditions under which scientists operated in the postwar period were not onerous enough, the heightening of McCarthyism, combined with the conviction that the Soviets were only a classified document away from nuclear parity, or even superiority, prompted an overzealous campaign to root out atomic spies in research labs throughout the country. This campaign, for many, signaled the ultimate triumph of irrationalism. A few celebrated cases of atomic espionage early on, highlighted by the arrest of Klaus Fuchs, provided the justification the various government agencies and committees needed to pursue spies at any cost;

the political climate of the time provided the incentive. They all came up empty-handed. The lack of bona fide atomic spies among American scientists, however, did not stop the House Un-American Activities Committee, for example, from recklessly interrogating witnesses and repeatedly casting aspersions on the loyalty of numerous scientists and technicians.[55]

All of this fanaticism, if not simply pretense for political attack, was driven by a public belief that there indeed was some "atomic secret" that could be stolen or given away. This was further evidence that the public found it difficult to distinguish science from technology, or patient inquiry into natural processes from mysticism. Repeatedly scientists tried to tell those who would listen that "in science, the true bearer of secrets is Nature, and no security policy can prohibit it from disclosing these secrets from others who are qualified to interrogate it."[56] Other countries, one physicist explained, already had all the information they needed: "When one is so fortunate as to know where to look and what results to expect, then discovery is certain. We can only conclude that the major secret of the atom was lost at Hiroshima on August 7 [sic], 1945, and that subsequent events will not modify another nation's time scale very greatly."[57] Many used this reasoning to argue for even more government support with even less interference. "In science," another scientist explained, "you cannot *keep a secret*. The best and most you can hope to do is to *keep ahead*."[58] Unfortunately, arguments such as these failed to persuade, and the misguided belief in atomic secrets persisted.

Scientists were not insensitive to the government's need to safeguard defense-related technical information at some level. It was the method by which it was accomplished that so appalled the scientific community.[59] To many, the loyalty investigations were handled in a way that seemed "at times arbitrary and capricious, motivated partly by politics."[60] Karl Compton, the former president of MIT, expressed his worry over "the detrimental effects, on our work for technological preparedness, of publicity regarding alleged espionage and the charges which have been made against reputable scientists on the basis of hearsay and unsubstantial evidence."[61] An analysis of investigation procedures completed by a committee of the AAAS concluded among other things that due process rights were frequently ignored, that no clear standards existed for making decisions regarding security clearances, and that charges were often vague and relied on secret testimony contained in FBI files to which individuals under investigation had no access.[62]

Most troubling of all, perhaps, was the apparent existence of an American "party line" to which all were expected to adhere. This brand of "thought control," offensive to most academics, was particularly distasteful to the scientific community, steeped as it was in the commitment to free inquiry.[63] "It is quite unnecessary to remind you of the many attacks upon scientists in recent years, that serve as straws to show the direction in which some of the anti-scientific winds are blowing," stated Kirtley Mather before a meeting of the AAAS. "If you don't like what a scientist says about the necessity for freedom from thought control in America, the wisdom of supporting the United Nations . . . or anything else that he may say, the easiest way to slap him down is to insinuate that he is dangerously subversive and advocates doctrines that are approved by the Communists."[64] To flout the Cold War political orthodoxy of the time was to court political reprisals, especially if one occupied a sensitive position in the national security state.

Few security cases touched the scientific community as deeply as that of Robert Oppenheimer. As the intellectual leader of the atomic bomb project at Los Alamos, Oppenheimer had served with unparalleled distinction. It was, thus, with shock and dismay that people greeted the news in December 1953 that the Atomic Energy Commission (AEC) had identified him as a security risk. Oppenheimer's affiliations with communist and other left-wing organizations throughout the thirties and early forties were well known. He had been authorized nonetheless to lead the Los Alamos group during the war and was approved yet again in 1947 to serve as chairman of the General Advisory Committee of the AEC. By the early 1950s, though, the political mood had shifted, and he was called to account for his past once again. The hearings before the personnel security board of the AEC in the spring of 1954 concluded with the board announcing its decision to revoke his clearance.[65]

It was evident to many that Oppenheimer had been targeted by officials in the AEC for his political views, especially for his stand in favor of the development of smaller tactical nuclear weapons over the development of the hydrogen bomb. Against Oppenheimer's and others' advice, the "super," as scientists called it, received the green light and soon became a high priority government project. During the AEC security hearings, government attorneys cast Oppenheimer's opposition to this established government policy as a conscious effort to subvert American nuclear superiority. That one's professional opinion as a member of an advisory committee could be twisted to suggest some treasonous intent infuriated scientists across the country. It especially angered the members of the elite group of physical scientists that had come to dominate the science advisory boards charged with ensuring American technological superiority.[66] The Council of the American Physical Society publicly stated that "charges based on policy disagreement appear to be customary in

Russia but we regard them as not only morally reprehensible but also very harmful to our national welfare." The Council concluded that "to require . . . subservience to an official viewpoint as a proof of trustworthiness is to prevent the development of the best thought."[67]

The Oppenheimer affair damaged the already fragile relationship between the government and the scientific community. Many believed it was time to speak out. Vannevar Bush, the dean of wartime science, laid out his concerns in a lengthy article in the *New York Times Magazine* entitled "If We Alienate Our Scientists." In measured tones he lectured readers on the threat to science and ultimately national security from the widespread public suspicion. Some security precautions were needed, he agreed, but "there is an enormous difference between taking due care and striking blindly in a wave of hysteria." The attitude toward scientists needed to change, he argued, they "need to be used not as lackeys or underlings but as partners in a great endeavor to preserve our freedoms."[68] The article was well received in the scientific community. Cornell physicist Hans Bethe commented in a personal note to Bush that the piece "expressed so well and so clearly the thoughts of us all."[69]

Others were less measured in their condemnation of the overall treatment of scientists. To many, it was not too much to characterize the behavior of the public and government officials as a form of madness. In a speech before the AAAS on the topic of the "freedom of thought and expression," the physiologist Maurice Visscher declared, "I think we live in a psychotic world. . . . The really insane people, by my definition are found in the vociferous minority groups, and the great masses of people are putty in their hands."[70] What so angered Visscher and so many others were the techniques deployed to quash any political unorthodoxy exhibited by scientists—techniques that struck at the very heart of scientific thinking. "In our insane desire to remove the imaginary threat to our security offered by the imaginary perfidy of our scientists," he lamented, "we are adopting the worst features of a totalitarian state and are stultifying our scientific free enterprise."[71] Clearly by 1954 scientists had found themselves in an environment they perceived as hostile to the best interests of science.

## The Advancement of Science

Though scientists by this time belonged to what the editor of the *Bulletin of the Atomic Scientists* referred to as "a harassed profession, occupying a defensive position in the political arena," they were not without either the means or the desire to redefine the public image of science along lines more favorable to their collective research agenda.[72] Through a series of initiatives, some coordinated at the national level and others undertaken locally, scientists began their effort to roll back the creeping public irrationalism and to integrate a renewed scientific outlook into the fabric of daily life. The Cold War ideological mood of the early 1950s, which culminated in the excesses of the Red Scare, was an important factor that not only prompted the scientific community to action, but also informed the image of science that it subsequently presented to the public—an image designed to serve the professional needs of the research community. The innumerable assaults on democratic freedoms underwritten by the paranoid fear of communist expansion were clearly objectionable to the scientists. As objectionable as they were, though, scientists recognized that the pervasive fear on which these assaults were based was deeply ingrained in the culture of 1950s America. Thus even as scientists sought to communicate what they saw as the true nature of their work to the greater public, they did so in ways that would comport well with the Cold War political orthodoxy of the time.

In examining these efforts to remake the public image of science, it is important to reemphasize the differences between the conceptions of science held by scientists and the nonscientific public. Although scientists, as well as other intellectuals, were viewed as rather odd at times and potentially subversive, most lay people held science itself in high regard. It was the exemplar of objective thought and the surest means for arriving at truth about the natural world. True as that was, the justification for its generous public support derived primarily from its technological accomplishments. For those preoccupied with national security, this meant the development of the "super" and the refinement of radar surveillance techniques. For the average consumer such accomplishments translated into, as one scientist described, "some new chemical with miraculous tricks, which will render gasoline ten times more powerful than before or will improve one brand of tooth paste much above another."[73] For many, science and technology were viewed as one and the same. Furthermore, this science as technology found greatest public support in the relatively limited domains of the physical and biomedical sciences, as the Cold War with the Soviets and the war on cancer and other diseases demonstrated. Venturing into the social realm often generated resistance from vocal conservative minorities, as was evident in the case of government-mandated fluoridation and the movement toward school curriculum grounded in theories of psychological adjustment.

This view of science was far from that held by members of the scientific community. Most would have agreed with the physicist Victor Weisskopf, who saw science as "the organized expression of the human trend to penetrate, clarify, and understand the world around us"—certainly a broad view not limited by concerns of material need or field of application.[74] Scientists, however, were not so naive as to believe that simply working to understand the natural world in and of itself had contributed much to their ascent to key advisory positions within the federal government, or that it had helped generate the financial support they currently enjoyed. As the American Association of Scientific Workers (AASW) observed in a position paper on the state of science, "during the past ten years science in the United States . . . has won for itself a public acceptance and support that it has never known before." It was, however, "fairly apparent at the outset . . . that science was not being supported for its own sake, nor for what it could do, but, in far too narrow and specific a way, only for those things that it had so spectacularly done." This, the AASW argued, was the ultimate source of nearly all the dangers now confronting science. The research teams at Los Alamos, Johns Hopkins, and the Radiation Lab had done their jobs too well. The success of the radical new weapons and military technologies set a precedent against which subsequent scientific work was measured. The result was that funding agencies and donors now "demand, without delay, further sensational successes of the kind to which the nonscientific world had become accustomed. No allowance is made for the time-consuming, broadly exploratory work that underlay the earlier successes. Again science is the victim of an oversimplified image of itself in the public mind."[75] This distorted image of science, many in the research community believed, lay at the heart of the challenges science faced in the years following the war.

Public misunderstanding of science was a theme that filled countless papers and talks in the scientific press. National Science Foundation director Alan Waterman in a 1955 *Scientific Monthly* article reiterated the widely held feeling of "danger that society may be moving either toward a rejection of science or toward an active control of its direction." The situation, he felt, was serious. "The nature of the scientific advances of the past decade, coupled with the political idiosyncrasies of our times," he argued, "makes the understanding of the advancement of science a practical matter of survival."[76] But the science of which Waterman was speaking was not that which would develop new weapons, but rather the pure science devoted to understanding the processes of the natural world.

Work clearly needed to be done to rehabilitate this image of basic science and to demonstrate its value not just as means to greater material welfare, but more broadly as a means to rational understanding—an understanding that, scientists believed, would greatly ameliorate the vast and imposing social and political problems that confronted the United

States.[77] Such goals depended first and foremost on public support of the scientific community. Biologist Marston Bates, who later became a key figure in the BSCS curriculum project, complained of the lack of effort scientists had put forth in generating this support. "We . . . have failed miserably in this, and we ought to face up to it," he declared. "If science itself hasn't got the intelligence to manage its relations with the cultural environment in which it has grown, maybe it deserves the legislative axe and the indifference of the public."[78]

Early localized efforts to remedy the situation included curriculum reform at the university level. Right after the war, Harvard president James Conant, having witnessed "the bewilderment of lawyers, business men, writers, public servants (and not a few Army and Navy officers) when confronted with matters of policy involving scientific matters," initiated an experiment in teaching science as part of Harvard's new undergraduate general education program.[79] The most innovative of the program's science courses was Natural Science 4, "Research Patterns in Physical Science," a course that Conant himself had developed and initially taught. Intended for those not pursuing further scientific training, the course used a series of historical case studies to illustrate "the ways by which the experimental sciences have advanced in the past and the methods by which they are still progressing."[80] The goal of Nat. Sci. 4, as it was called, along with the other courses in the program, which covered more standard topics in the history and philosophy of science, was to help students understand the essence of scientific research. The hope was that a proper understanding would lead to an appreciation of the complex relationship between science and society and would thus ensure appropriate levels of research autonomy and public support. Similar courses were developed in the 1940s at the University of Chicago and Columbia University targeted to the intellectual elite of the country as part of the broader postwar general education movement.[81]

At the national level, the AAAS made one of the first systematic attempts to address the growing threats to science head on in the fall of 1951. In response to a rather blunt challenge from a former association president to take up the defense of science, the executive committee convened a special meeting to reexamine the role of the AAAS in supporting scientific work. There evidently was much to be concerned about. "Contrary to statements we have heard we do not believe that this is an age of science. Rather this is an age in which a few scientists live among primitive men possessing the products of scientific endeavor," observed three of the consultants to the meeting, reflecting the progressively deteriorating conditions under which scientists labored. "The concepts of free inquiry and non-authoritarianism, of integrity and pragmatism—to mention a few principles of scientific activity—must be conveyed to the public," they went on, if the future of science is to be secured.[82] The summary of that meeting, the "Arden House Statement," named after the Columbia University host facility, charted a new course for the association that emphasized the health and progress of the scientific profession as a whole.[83] To this end, the AAAS board of directors established a committee to address "the broad external problem of the relations between science and society" by focusing on increasing "public understanding and appreciation of the importance and promise of the methods of science in human progress."[84] Though it had some difficulty getting started, the committee eventually sponsored programs to improve science journalism and teaching, and even produced popular books, and radio and television programs that covered important scientific issues.[85]

The movement to further the interests of science initiated with the Arden House Statement was reinforced two years later by an international conference on "Science and Freedom" held in the West German city of Hamburg. Located a stone's throw from the communist eastern sector, the conference sought, as its primary goal, to "summon the attention of the world to the damage done to science by totalitarianism."[86] Along with condemning the communist regimes of Europe, the conference participants condemned the oppressive features of the loyalty and security programs of the United States as well.[87] Totalitarian tactics, they argued, threatened science regardless of their country of origin.[88]

The most telling morality play regarding the dangers of state interference in science had been staged on the red soil of the Soviet Union. According to communist ideologues, the radical theory of knowledge, dialectical materialism, on which the Soviet political system was based, enabled a scientific understanding, even prediction, of the direction of human progress. To followers of this view, it seemed that the inexorable social progress predicted by Marxist theory could be facilitated by the careful scientific management of Soviet society.[89] In this technocratic culture, technological progress was synonymous with social progress. To this end, Soviet science collapsed the distinction between theory and practice; in the philosophical materialism of Marxism, understanding was inseparable from action. As the Marxist scientist J. D. Bernal explained, "the appreciation of the social position of science leads at once in a socialist country . . . to the organic connection of scientific research with the development of socialized industry and human culture."[90] In essence, all science became technological development in the service of the state. Basic science had no place in the properly constituted scientific society.[91]

As Soviet scientific practice ran afoul of Marxist theory, it was soon obvious to Western scientists that state intervention could have disastrous effects. Such was the case when the well-established theory of Mendelian genetics was deemed by Soviet authorities to be incompatible with Marxist philosophy. The ideological bias against theoretical research programs like that of classical genetics combined with the long-discredited mechanism of the inheritance of acquired characteristics advanced by the politically savvy agronomist Trofim Lysenko devastated Soviet research in genetics and severely handicapped the country's agricultural productivity. The subsequent purges of dissident geneticists outraged scientists worldwide and confirmed for many the destructive power of ideology on science. H. J. Muller, an American geneticist who had worked in the Soviet Union from 1933 to 1937, upon his return to the United States conceded the failure of state-run science in this case. "We must confess that we no longer see any chance of saving the core of biological science, and all that goes with it, in that section of the world in our generation, short of an unexpected political overturn," he stated. The only hope was to check "the already dangerous spread of the present infection to countries outside of the Soviet hemisphere."[92]

The Hamburg conference was emphatic in its desire "not only to denounce the harassment and deformation of science in the Soviet Union but also to make Western scientists and scholars more fully aware of what is entailed in the claim to freedom for the pursuit of truth."[93] The conference proceedings highlighted and reinforced the emerging view of the central feature of science in the 1950s: its reliance on the complete freedom of inquiry guided by nothing greater than the rational appraisal of the bare facts of nature. Just as membership in the Communist Party required lock-step adherence to the ideological party line, as Sidney Hook argued in his stand against communist teachers, membership in the community of scientists at the other extreme committed one to the pursuit of truth, which entailed the conscious rejection of all ideology. As one conference participant explained, "the very act of participating in scientific work and accepting membership in the scientific community is intrinsically incompatible with dogma. The strength of the tradition of scientific life and scientific institutions reaffirms itself against the claims of ideology."[94]

## Science and Democracy

More important in the public assessment of the nature of science than its incompatibility with totalitarian dogma was its growing identification with democracy.[95] As intellectuals struggled over the philosophical foundations of the seemingly fragile democratic system of government following the Depression, they increasingly invoked science, with its powerful rationalism and legitimacy, as the model on which democracy was ultimately based. Democracy and science, as philosopher John Dewey had argued for decades, were really just two forms of the same worldview. "Freedom of inquiry, toleration of diverse views, freedom of communication, the distribution of what is found out to every individual as the ultimate intellectual consumer," he explained, "are involved in the democratic as in the scientific method."[96] As the ideological threat from the Soviets mounted in the early 1950s at home and abroad, scientists argued ever more convincingly that the dissemination of scientific thinking was one sure way of promoting democratic ideals. "The best antidote to communism, which is hampered by its attachment to an out-dated doctrinaire Marxism," observed one political historian of the time, "is a more rational and effective democracy, linked to the actualist methods and spirit of science."[97] A scientific attitude was also seen by scientists as a positive means to counter the antidemocratic tactics of McCarthy and other red hunters in the United States.[98]

The democratic foundations of science, scientists forcefully pointed out, however, were characteristic only of pure science. They seized on this close association with democracy to push hard for greater freedom in the organization and pursuit of scientific research. If totalitarian ideologies were inimical to science, they argued, then it was clear that external interference of any kind would be detrimental to scientific progress as well. The freedoms inherent in democratic social arrangements must be secured for science if it is to function at its optimal level. As participants of the Hamburg conference recognized, this raised the vital question of "how the freedom of science could be guaranteed when government subvention was necessary"—the question again of the tension between patronage and autonomy. The scientific community could offer only reassurance that funding without excessive government control was, over the long run, in the best interests of society. "One very valuable contribution of the whole conference," sociologist Edward Shils explained in his summary of the meeting, "was the gradually developed notion that scientific activities . . . formed a kind of social and cultural system with its own powers of self-maintenance and self-regulation and that this system must of necessity be relatively autonomous."[99] Autonomy in the pursuit of basic research, as Vannevar Bush had argued in his call for a national science foundation, would ultimately benefit all of society in the relief of material wants and maintenance of national security.

Although American scientists in the early 1950s might have been frustrated by their wait for the arrival of their "scientific age," many could take comfort in the fact that they were making some progress at least in getting their message out. To those who challenged the value of a society grounded on a scientific worldview, scientists reasserted the power of rational thought not only as the engine of material and social progress, but also as the only means to combat the rise of totalitarian ideologies.[100] To others who sought to bind science to the service of the state, scientists pointed to the disaster that befell Soviet biology and reemphasized the crucial distinction between pure and applied research, between basic science and technology, and with shared voice insisted on the freedom necessary for the successful pursuit of basic scientific inquiry. For those in Congress wary of appropriating funds over which they would have little control, as David Hollinger has convincingly argued, scientists presented themselves as members of "a particularly virtuous community possessed of distinctive interests." Those interests included a commitment to the pursuit of truth free from ideological distortion, a belief in the sharing and dissemination of knowledge, and a social organization that was inherently democratic. Given these characteristics, and the government's nearly total dependence on scientific expertise for national security, "it was all the more legitimate, then, to allow such a community to have autonomy within a democratic community like the United States."[101]

The effort invested by the various advocates of science paid minor dividends beginning in the mid-1950s. The fortunes of the scientific community began to rise partly in response to the public relations work of groups such as the AAAS and partly as a result of the relaxation of domestic tensions following the resolution of the Korean crisis and the waning of McCarthyism. But scientists were keenly aware that rapid reversals could still take place. In his 1956 retrospective on the security hysteria in the United States, Shils opened with the observation that "after nearly a decade of degrading agitation . . . the disturbance aroused in the United States by the preoccupation with secrecy and subversion has begun to abate." He explained, however, that "this abatement should provide no occasion for self-congratulation."[102] On the contrary, the clear message for scientists was that they needed to be ever vigilant in the protection of their practice. Indeed, they would find, by the late 1950s, with the rush to harvest the technological fruits of science following the Soviet launch of *Sputnik,* their situation not appreciably better.

As maligned as scientists felt during the height of the Red Scare, science itself, even viewed primarily as a means to technological mastery, retained its long-standing connection to truth and rationality, particularly in its method of inquiry. It thus proved to be an attractive ally in Arthur Bestor's attempt to establish a national committee to recenter the academic disciplines in the school curriculum. Given the growing reliance on technology for national security along with the renewed emphasis on the virtues and characteristics of basic science in the public discourse, greater scientific leadership in the troubled and ideologically strained arena of public schooling in the spirit of the general education movement at the college level seemed increasingly assured. True to their commitment to rational thought, nearly all scientists professed a vague faith in scientific education as a means to ameliorate the problems of postwar society. More pressing for a group recently awakened to its vulnerable position in a democratic society was the problem of ensuring public support for a professional scientific community practicing science on its own terms, well funded and safely shielded from public interference. Education was the means to that end as well. The only questions remaining were when and in what form would scientists enter the classroom.

## CHAPTER 3

# NSF, EDUCATION,
# AND NATIONAL SECURITY

E ven with the scientific community's keen interest in education as a means
to correct the public image of their work, and the clearly evident inadequacies of American education brought to light in the early 1950s, few scientists considered the public schools a viable forum for disseminating their
view of science. The decentralized organization of the American school system made any aspirations for comprehensive reform, along whatever lines,
difficult to envision. The highly charged political atmosphere that attended
the early curriculum debates over life-adjustment education, moreover, made
involvement in education at the pre-college level even less attractive. From
the federal government's point of view, issues surrounding the quality and
aims of schooling were best left where they had always been, safely in the
hands of local communities. In a 1953 report drafted by the U.S. Commissioner of Education for the incoming Eisenhower administration listing the
"major educational issues currently facing the nation," not one item concerned the quality of instruction at the elementary or secondary school level.[1]

Federal interest in this topic, however, particularly with respect to science education, was to pick up considerably during Eisenhower's two
terms as chief executive. The technological race with the Soviets soon provided all the incentive necessary to prompt a reconsideration of American
educational policy. Reports out of the Soviet Union describing their centralized, highly efficient system of scientific and professional training, combined with the scientific manpower shortages in American government
research labs and industry, pushed science education, at all levels, toward
the top of the federal agenda by the middle of the decade. Numerous questions remained regarding the specific nature of the problem and the best
course of action to take. These still needed to be resolved, and the path of

their resolution was dependent on a number of institutional and political factors. All eventually pointed to the National Science Foundation as the most logical agent of change.

As the government agency charged with ensuring the nation's scientific strength, the National Science Foundation found itself pressured from the top by influential members of Congress and various government committees to address programmatically the pressing issue of ensuring an adequate supply of scientific talent to meet the mounting Soviet threat. The young and underfunded Foundation, fully occupied with its self-described task of supporting basic research, was wary of entering the politically explosive field of pre-college education. The highly dispersed structure of American schooling remained a formidable obstacle to overcome as did political opposition to federal interference in local school affairs. In the end, however, the National Science Foundation moved decisively to reform high school science education.

The kind of programs Foundation officials sponsored, what they chose to address and what to ignore were strategically developed to generate as widespread an effect as possible given limited resources and existing institutional barriers. Perhaps the most impenetrable of these was the largely self-contained system of teacher education and state certification excoriated by Bestor. Few officials who studied the problem of inadequate teaching felt that the federal government could have any real leverage in changing the patterns of pre-service teacher training. The only realistic option was to engage existing classroom teachers directly. Initially this took the form of summer training to update their deficient science content knowledge, which was followed by the production of much needed curricula. Such initiatives required not only greater government appropriations, but hard congressional support as well. This support was indeed forthcoming, primarily in response to the evidence of a strong Soviet push in high school science education. That, along with the long-standing desire of many in Congress to provide federal aid to schools and the perception of scientific knowledge as nonideological, made support of the NSF programs politically palatable. The fact that the professional education establishment was excluded, despite evidence of public support of their functional curricular ideology, demonstrates the overriding influence of both national security and the scientific elite in redefining the school curriculum in the 1950s.

## The Problem of Scientific Manpower

The manpower crisis that erupted across industry, government, and academia with the outbreak of the Korean War powerfully illustrated for those directing the research enterprises of the United States just how valuable a commodity scientific expertise and technical proficiency had become since World War II. Just when it was thought that the "'shortage-of-scientists' problem . . . [was] about over," the American military action in Korea generated an immediate expansion of scientific research and development to meet the increased needs of the defense department during this national emergency.[2] This military-fueled R&D expansion was led by the aircraft, electronics, and machine tool industries. The electronics industry, paced by companies such as General Electric, Raytheon, and Westinghouse, for the most part owed its very existence to defense department contracts and depended heavily on scientific manpower to fill them.[3] The existing supply fell short. The problem, as the National Science Foundation observed, was that the growing demands for scientists in the 1950s had met "the 'thin generation'—the small age groups born in the 1930s."[4] The importance of this demographic problem to the industrial and academic scientific elite was highlighted by the publication of a special issue of *Scientific American* in September 1951 devoted entirely to a discussion of the human resource problems of the United States. The issue sold out immediately and was reprinted three times before demand finally slackened.[5] Despite the government's longstanding laissez-faire approach to education, the Eisenhower administration had come to view the shortage of scientists as a serious threat to national security, thus prompting action that put the federal government on the path to its unprecedented involvement in pre-college science education.

Initial efforts to alleviate the shortage centered on finding short-term solutions. More scientists could not simply be found; they needed to be trained. It was well known that a "physicist or chemist cannot be turned into a creative and productive worker in less than five years."[6] Given the likely timetable of the Korean crisis, manpower experts focused their energy on the tail end of that rather long production pipeline. The National Science Foundation channeled its limited resources into providing graduate fellowships for advanced scientific training. The selective service system did its part by granting draft deferments to college undergraduates, viewed as the pool from which future scientists were drawn, on the basis of test scores and class standing. Science instruction at the high school level (the very beginning of the training pipeline) was largely ignored by government agencies. Despite the growing shortage of teachers across the country in all subject areas, the editors of *Scientific American* saw fit to include only engineers, scientists, and doctors as occupations of concern in its special issue in 1951, deciding perhaps that the problems of the public schools were too far removed from the crisis at hand.[7]

One interesting consequence of the manpower crisis was the attention it focused on the role of women in scientific and technical fields. An early editorial in the *Bulletin of the Atomic Scientists* identified one "large reservoir of potential scientific, medical, and engineering 'manpower,' which [was] as yet almost entirely untapped: American women."[8] The millions of girls already in college, none of whom could then be drafted, the editorial argued, merely had to be encouraged to pursue science over studies in the arts and home economics. Arthur Flemming, the director of the Office of Defense Mobilization (ODM) also insisted that "women should not be overlooked" as a potential solution to the manpower problems facing the nation.[9] Unfortunately, despite repeated encouragement, occupational opportunities, even for those well-trained in the sciences, remained limited. Many of those writing about the manpower problem saw women contributing to the solution only by filling the innumerable routine "jobs to be done in hospitals, laboratories, and industrial shops," in order to free up men to fill the "positions of leadership."[10] Later in the decade, when the shortage of high school science teachers was viewed as critical to national security, manpower experts argued that increasing the number of women studying science might help resolve that shortage since they were more likely than men to enter and remain in the teaching profession.[11]

While most organizations during the crisis placed their emphasis on immediate short-term needs, some possessed enough foresight, even before the Korean War, to invest in long-range scientific training programs. As early as 1945 General Electric established a six-week summer school experience for high school science teachers at Union College in Schenectady, New York. The stated goal of the program was, straightforwardly enough, to "improve . . . the teaching of science in the high schools of the country."[12] The first summer institute, as they came to be called, was in physics and covered current topics such as modern physical theory; electronics, including vacuum tube theory; and modern applications of physical measurements. The curricular emphasis was exclusively on science content, making no attempt to address issues related to pedagogy. The instructional format was traditional—strictly formal lecture and laboratory work. This rather demanding program was sold to prospective participants as a means to acquire cutting-edge knowledge in modern physics, and, if that were not enough, there was the promise of personal contact with big-name GE scientists such as Nobel laureate Irving Langmuir. Whatever the reasons, there appeared to be plenty of demand as hundreds applied for the first 40 slots.[13]

General Electric, of course, was not completely magnanimous in providing this training for teachers. Certainly GE corporate officials believed that better science teachers were essential for increasing the number of high school students pursuing science in college and thus would help bolster the scientific manpower on which they depended. But the summer program also served the more self-interested goal of indirectly bringing the "General Electric Company to the attention of students in the high schools of the country."[14] To this end the company sprinkled mandatory screenings of GE public relations films and tours of nearby GE facilities throughout the six-week program. GE expanded the institutes to include teachers of chemistry the following year and eventually sponsored training in various technical subjects at three other college campuses across the country. Westinghouse, which had earlier established the famous science talent searches, followed GE's lead and started up its own summer institute program in cooperation with faculty at MIT in 1949. These early programs, however, were novelties at best in the larger scientific manpower picture. Nevertheless, they did provide an early model for addressing some of the shortcomings of the nation's pre-college educational infrastructure.

Whether long term or short term, the tactics employed to alleviate shortages of scientific personnel had only limited success. Demands for greater attention to the problem grew insistent as the Korean War progressed. Industrial and professional scientific organizations had, for some time, generated a variety of advisory reports for the government. Professional journals circulated among the scientific-industrial elite all devoted space to this pressing issue.[15] Among these, the consensus seemed to be that national survival depended on increasing the pool of scientific personnel, which was crucial to staffing the record numbers of projects undertaken by the defense industry. Thus corporate prosperity and national security made common cause. News of this pressing need, however, failed to travel far beyond the desks of corporate research directors. The one group that appeared to succeed finally in drawing public attention to the problem was the National Manpower Council.

Established in 1951 at Columbia University, the National Manpower Council was an independent commission set up with a grant from the Ford Foundation. Council membership included individuals from the Department of Defense, leading research universities, technical institutes, and various industrial corporations. The group's objective was to "appraise continually the problems and policies bearing upon manpower utilization and to analyze and report thereon to policy makers and to the public." This it accomplished through the publication of a series of reports over the course of its existence.[16] Dedicated to promoting what the council saw as the limited "public understanding of the crucial role 'brainpower' plays in the nation's economic progress and military defense," they achieved a breakthrough of sorts with the publication of their report entitled *A Policy for Scientific and Professional Manpower* in 1953.[17] Newspapers across the country featured the report and its conclusions, hailing it as a landmark in "directing public attention to a neglected field." Coverage extended to all the major radio and television networks as well as popular magazines and even to some foreign newspapers.[18] Ironically, only months after the report's publication, the newly installed Eisenhower administration agreed to a United Nations brokered armistice on the Korean peninsula, effectively ending the war along with the immediate manpower crisis.

The importance of "brainpower" during the Cold War, however, was not lost on the new administration. Eisenhower, partly guided by his firsthand experience with personnel shortages during World War II, had been instrumental in establishing the Conservation of Human Resources Project as president of Columbia University in 1949, members of which later worked closely with the National Manpower Council.[19] Though the urgency that attended the issue of scientific training perhaps ended with the Korean War, the overall concern did not. The resulting peace, in fact, provided the necessary breathing room to undertake a more careful examination of scientific personnel needs. Soon after taking office, Eisenhower initiated a series of comprehensive reevaluations of government policy. Characteristic of his administration's approach was the careful coordination of all policy, foreign and domestic, to prepare the United States to endure and succeed in its Cold War battle with the communists.[20] The development of his "new look" foreign policy, with its emphasis on sophisticated defense systems, ensured that matters of scientific and technical training at home would not be neglected.

Directed action in this area, however, came sooner than many expected. Information from a top-secret CIA assessment of Soviet scientific training presented at a meeting of the National Security Council in October 1953 led the president to raise the issue of technical training at a cabinet meeting the following April. In both meetings, Eisenhower's advisors expressed concern that the United States might be outdone in science education and, perhaps, lose its technological and, therefore, military advantage. The CIA report began boldly with the statement: "The USSR is training a body of scientists and technicians which is increasing in size and quality and approaching comparability with that of the United States." The report specifically addressed the role of high schools in educating scientists, noting that in Soviet secondary schools there is a "far greater and more consistent emphasis on scientific subjects . . . than is found in the United States."[21] Cabinet members discussed how the federal government might work to encourage greater efforts in science education. It was clear that the subject called for more long-range planning. Robert Cutler, head of the National Security Council, urged that such planning proceed before the "situation becomes more critical with the passage of time." At the end of the meeting, Eisenhower directed ODM head Arthur Flemming to establish an informal committee to look into the matter in greater detail.[22]

Less than a month later, Flemming contacted National Science Foundation director Alan Waterman informing him of the formation of the new Special Interdepartmental Committee on the Training of Scientists and Engineers (SICTSE). NSF was to play a central role along with representatives from the Atomic Energy Commission and the departments of Defense, Labor, and Health, Education, and Welfare in evaluating the status of U.S. science education. The ODM, which led the push for federal involvement, was clearly concerned about the shortage of scientists. "If this were merely a temporary lag," an ODM internal memo stated, "the matter would be of no more than general concern to the federal government." It went on, "today, however, the security of the nation is at stake. The danger lies in our possible inability to maintain the present substantial margin of superiority in scientific and technological development which we have over our potential enemies."[23] With the establishment of SICTSE, science education became another tool in America's Cold War arsenal.[24]

### Targeting Science Teachers

The new look at science education came as the country had begun to settle into an uneasy period of relative quiet. Though nearly a year had passed since the end of the Korean War, and McCarthyism was rapidly fading from view, Eisenhower urged the country not to become complacent. The Cold War was far from over. Indeed, he noted at a news conference that it might persist for another 40 years, a point seized upon by the press in its coverage.[25] The longer the conflict was predicted to last, the more important education seemed to become, and the more troubled people were by the existing state of the American education system. As the president's interdepartmental committee and the National Science Foundation surveyed the educational landscape, it became clear that the problem, just as with the country's research infrastructure, was one of inadequate manpower resources. In this case, it was the lack of qualified teachers—especially in the sciences—that needed to be addressed.

Eisenhower's formation of SICTSE generated immediate friction with the young but already established National Science Foundation. Since its inception, science education and training had been one of its primary charges. To oversee this part of its work, the Foundation had established the

Divisional Committee for Scientific Personnel and Education headed by Harry C. Kelly, a physicist who had previously worked for the Office of Naval Research in Chicago and at MIT's Rad Lab during the war.[26] Kelly was well aware of the importance of scientific manpower to the defense industries and his committee had worked hard, if cautiously, to develop a program to meet the country's needs in this area. NSF officials thus viewed Flemming's interdepartmental committee as at best unnecessary and at worst interfering in its own work in education.

Apart from this concern over intrusion in NSF affairs was the more substantive concern over the kind of programs that might be initiated to address the nationwide manpower shortage. Here the biggest difference could be traced to the constituencies each group represented. SICTSE was a product of the Office of Defense Mobilization and was thus primarily concerned with meeting the demand for scientific personnel in defense industries. Its focus was on increasing the absolute numbers of scientists and engineers coming out of the training pipeline.[27] Waterman and the rest of the NSF administration, who were largely drawn from the nation's scientific elite, argued that a focus on numbers alone would serve only to dilute the quality of the scientists produced. Foundation officials felt that the shortages, for the most part, were overstated, being really acute only in mathematics and physics. Even in those fields they believed that, simply given some time, supply and demand would likely even out.[28] What was needed, they argued, was a steady long-range policy, rather than some ill-conceived crash program that would do more harm than good.[29]

Despite this disagreement over the relative urgency of the manpower problem, there was a consensus among both NSF and ODM officials over its ultimate source: high school science teachers. Some members of NSF's Divisional Committee for Scientific Personnel and Education had felt all along that "poor teaching [was] the heart of the problem."[30] At the first meeting of the Special Interdepartmental Committee, this point was raised again by a representative from the Atomic Energy Commission. In response, Waterman commented that "all who had studied the problem felt the same way that, namely, the crux of the problem lay in the secondary schools and there the chief difficulty was in securing competent teachers of science."[31] Evident in Waterman's statement was the twofold nature of the problem: most apparent was the raw shortage of teachers; less obvious, though more important from NSF's point of view, was the low quality of those currently in the classroom.

The teacher shortage was a chronic problem obvious to anyone who kept abreast of current events in the early 1950s. Though not highlighted by the editors of *Scientific American* in 1951, the National Manpower

Council by 1953 had explicitly tied this problem to the crisis in scientific manpower production. Their breakthrough report of that year devoted an entire chapter to an evaluation of the supply and demand of public school teachers. The elementary schools, the report noted, experienced the most pressing shortage. However, as numerous other studies had established, the problem would soon overwhelm secondary schools as well. Discussions at NSF noted that between 1950 and 1965 three hundred thousand more secondary school teachers would be required. In science subjects alone that translated into a need for up to ten thousand new science teachers per year over that time period. And this came at a time when the absolute numbers of college graduates were declining.[32]

The variety of factors affecting supply and demand were difficult to pinpoint, much less control. One situation that obviously discouraged many capable individuals from entering the profession was the low pay.[33] In the absence of additional state or federal aid, local communities, already over-burdened in many cases with expensive building projects, were hard pressed to raise salaries. Further contributing to the science teacher shortage, in Waterman's eyes, was the low public esteem such positions held in the United States. In comparison he noted, "Russia has an enormous advantage in the prestige which she attaches to science and the importance she attaches to the training of scientists and engineers." In this regard, he went on, "we are behind other countries, especially with respect to our science teachers in secondary schools."[34] The problem seemed far beyond the reach of either SICTSE or NSF. Both recognized that, except for drawing attention to the problem and encouraging public support of teachers, little could be done.

The problem of teacher quality could not easily be separated from the problem of numbers. The sheer inability to find certified teachers to fill the growing number of classrooms led to a variety of stop-gap employment practices. Certified teachers were often pressed into service teaching subjects in which they were not qualified. Or worse, individuals with no educational training whatsoever found themselves leading classrooms across the country.[35] In 1953, nearly 65,000 teachers were working with emergency certification, a classification that was essentially a waiver of any requirements for the job. Concerns over quality, however, were not limited to those who lacked proper professional credentials. The National Manpower Council observed that certification was only a rough and, in many ways, questionable measure of teacher competence. The common perception of the time was that individuals who chose to go into teacher training programs had considerably less innate ability to start than other college students. This perception was reinforced by a survey of standardized test results from the Selective Service, which revealed that education majors as a group were only half as likely as engineering majors to score well enough to become eligible for draft deferments. Thus, it was generally accepted that the pool of students from which future teachers were then being drawn was of rather poor quality.[36]

What made matters worse, according to those occupied with the manpower problem, was the grossly inadequate training these students subsequently received. Teachers colleges and education programs at larger universities were rarely provided with the necessary resources to ensure top-notch preparation of their students. As the Council reported, "admission requirements, faculties, curricula, and libraries are below the standards prevailing in other colleges." Most suspect in all of this was the teacher-training curriculum itself. Professors of education, by now thoroughly maligned within the academy and publicly humiliated, were cited for teaching that lacked "depth and intellectual rigor" and requiring courses in educational methods at the expense of those in the traditional subject-matter areas.[37]

Members of the scientific community, just beginning to participate in the debate over public education, found this last practice of pushing coursework in teaching methods particularly damaging to the future of science. Many firmly believed that mastery of subject matter was absolutely necessary, and perhaps even sufficient, for good science teaching. "A teacher must know the subject of physics . . . in order to teach it, otherwise more harm is done than good," the physicist F. Wheeler Loomis declared at an NSF divisional committee meeting on the subject.[38] This view was seconded by many who were convinced that the average science teacher simply did not know much science. One GE institute director described the typical incoming teacher as "woefully lacking in a real comprehension of the fundamentals of physics. He has had little experience in problem solving and knows very little about atomic physics, nuclear physics, or the modern approach to solids. He may well have a masters degree, but 'Educators' and University physicists have forced him to take this in 'Education.'"[39] This evident inattention to subject matter, combined with the explosion of scientific knowledge since World War II, left little doubt that science teachers were far removed from the cutting-edge of science.

One of the more frequently offered solutions to this problem was simply to encourage teacher-training programs to "pay more attention to grounding teachers in the substance of their subject"—a solution more easily suggested than implemented.[40] The various state teacher-certification requirements and general educational bureaucracy, which had so frustrated the Illinois historian Arthur Bestor, appeared to be an insurmountable obstacle to those seeking to emphasize science subject matter in teacher-training programs. The sentiment among scientists and other academics was that the professional educational establishment in this way contributed more to the problem of poor science teaching than to its solution.[41]

Given the magnitude and seemingly intractable nature of the teacher shortage problem, the members of SICTSE and NSF cast their attention instead on what could be done to improve the quality of science teaching already occurring in the nation's public school classrooms. This strategy allowed them to bypass the troublesome area of teacher-certification programs. Well before the formation of SICTSE, the Division of Scientific Personnel and Education at NSF had itself begun to explore ways to improve science teaching. Increasing teacher competence in modern science content was high on the list. In a 1952 memo, Kelly noted that "among the longer range undertakings . . . is the support of teachers of science; to bring them up-to-date on science through course study or to become reacquainted with modern research through participation in it." Though science faculty members of smaller colleges were the Foundation's primary target, high school science teachers were also mentioned as possible beneficiaries of such programs.[42] The first step in this direction came at the end of 1953 when the Foundation agreed to fund their first summer institute for high school teachers at the University of Washington. This institute in mathematics closely followed the format of the popular GE institutes and an earlier summer program for high school physics teachers held the previous year at the University of Minnesota, which was supported by the Fund for the Advancement of Education.[43]

In discussions held among members of the education division in the spring of 1954, the school curriculum emerged as a second key target area toward which the Foundation contemplated directing its resources. "One of the stubborn problems in science education," commented one committee member, "is the extent to which revision of the college curriculum, inevitably followed much later by the revision of the secondary school curriculum, lags far behind changes at the frontier of science."[44] The argument was that teachers, even those trained in cutting-edge science at the various summer institutes, were significantly handicapped in their teaching by the inadequate, out-of-date curricular materials they had at their disposal.[45] Talk of curriculum reform, however, immediately raised red flags in the minds of other NSF officials, who warned of the "dangers inherent in support by the Foundation of textbooks."[46] Waterman, ever protective of the fledgling foundation, wished to steer far clear of any public controversy that might threaten its continued funding. Foundation officials, in addition, continued to resist large-scale programs of any sort that might compromise the quality of scientific training and research simply for the sake of increasing the quantity of those trained. Though NSF had committed to supporting independent summer institutes for high school science teachers, it did so only cautiously and with much deliberation; taking on the public-school curriculum was another matter entirely.

The potential danger of mixing science with social issues was all too apparent to Waterman.[47] In the early congressional debates over the establishment of the Foundation, the prospect of including social science research in its charter raised numerous objections. One representative exclaimed that he would not support "an organization in which there would be a lot of short-haired women and long-haired men messing into everybody's personal affairs and lives."[48] The reactionary political temper of the times was not conducive to government-sponsored research that strayed far from the traditional subjects of scientific inquiry, which retained their aura of objectivity. Even nongovernmental research in the social realm was likely to invite undesirable public attention. From 1951 to 1954, officials from the leading private foundations, such as Ford and Rockefeller, that had ventured to examine pressing social issues found themselves testifying before congressional committees, charged with using their considerable resources to undermine traditional American values and abet the communist cause. The hostile reaction to the use of social science research in the 1954 Supreme Court *Brown v. Board of Education* decision to end racial segregation in public schools provided further evidence of the volatile nature of public opinion.[49] All these episodes testified to the strong, Cold War-inspired suspicions many had of any centralized authority, scientific or otherwise, interfering in their daily lives. As we have seen, schooling was an especially sensitive area in this regard. Waterman and other NSF officials had no doubt witnessed the widespread curriculum controversies of the early 1950s and were not eager to become entangled in a matter that might prove to be detrimental to the Foundation.

The ODM, however, had little patience for NSF's cautious approach. In its view, the Soviet technological threat to the United States far outweighed any self-interested concerns on the part of the Foundation to avoid political controversy. The informal interdepartmental committee led by Flemming of ODM had decided to recommend that the president establish a high-profile national committee to examine the nation's efforts in science education in order to draw much-needed public attention to this important issue. All the interested agencies in the administration gave the plan the green light. NSF was the one holdout.[50] It was also, however, the one government agency with the necessary expertise and statutory authority to do something about the problem. In an effort to get the Foundation on board, ODM officials sought to exploit public fears of communist expansion—an effective tool of the time. The message that ODM had begun to spread, that science was where "the struggle for military supremacy today is being waged," generated the necessary heat. In a letter to members of the Divisional Committee for Scientific Personnel and Education, Kelly described the mounting pressure for immediate action, which was coming "not only from the proposed report from ODM," but also has "been greatly heightened by the rash of newspaper and magazine articles on the subject."[51] In light of this, it was, as one education committee member noted, "probably desirable to 'cooperate with the inevitable.'"[52] Kelly duly recommended that the proposed program budget for science education for fiscal 1955 be increased from $500,000 to $2 million.[53]

## Mustering Public Support

The alignment of the National Science Foundation, in name if not spirit, with the scientific manpower plans of the Office of Defense Mobilization was only the beginning of a much more challenging task—that of convincing the country at large that American science education was in desperate need of reform. Any federal initiative in this area, however, if it had hopes of being funded, first had to find approval in Congress, long the jealous guardian of local control and hypersensitive to the prevailing public suspicion of centralized authority. Generating this broad level of popular support was no easy task. But the scientists had both the growing technological threat of the USSR and the perceived apolitical nature of science and the National Science Foundation as powerful allies to their cause.

One of the primary obstacles the Eisenhower administration faced in trying to address the shortage of scientific personnel was the absence of any systematic organization of schooling at the national level. American education was highly decentralized both administratively and politically. In this area, the enemy enjoyed a decided advantage. As the National Manpower Council observed, "Soviet Russia's current Five Year Plan calls for specified percentage increases in the numbers of natural scientists and engineers. In a totalitarian society such targets can be set, and all the power of a monolithic state can be employed to fulfill them."[54] This top-down, command-control organization of the Soviet system contrasted sharply with that of the United States, where the decision-making power was unevenly spread among "48 state governments and the school boards of some 59,000 school districts, as well as with a great variety of private and religious agencies." Given this organizational structure, it was obvious to everyone concerned that "changing programs in terms of the national interest cannot be achieved quickly or as efficiently as we might like."[55] In the technological competition with the Soviet Union, this fragmented system of schooling, by the lights of ODM, clearly threatened U.S. military superiority and cried out for some centralized coordination and control.[56] Though committee discussions and public pronouncements were sensitive to the inherent tension that existed between state needs and the democratic value of individual freedom, the immediate task at hand was to find the most efficient means to manipulate the existing educational system to produce more and higher quality scientists.

Despite deferential nods to local leaders in "the fields of education, the professions, industry and labor organizations," officials at ODM were convinced that the only viable course of action was to empower some federal agency to direct reform efforts.[57] Their obvious choice was the National Science Foundation, which had already begun down this path with its funding of the summer teacher-training institutes. Expanding these education programs, however, would require more than merely the cooperation of NSF. Congress, with its control over the appropriation of funds for the Foundation, would need convincing as well—even more so given its historical reluctance to venture into the quagmire of public school funding.

As mentioned previously, proposals for federal aid to education enjoyed no easy path on Capitol Hill. Since the end of World War II, general federal aid bills advocated by educational interest groups such as the NEA and the American Federation of Teachers made regular appearances before congressional committees with little success. With the onset of the school building crisis of the early 1950s, federal aid proponents changed their tack, calling on the federal government specifically for school construction funds to meet the growing demand for classroom space. Building needs were clearly evident across the country, and public demand for financial relief was high. The popularity of the issue made it essential, as Vice-President Nixon suggested at a cabinet meeting in 1955, that the president at least "be identified with a school construction program," whether implemented or not. Nearly every federal aid bill, however, became entangled in that intricate web of special interests known as Congress. Religious issues proved to be the downfall of aid bills during the Truman administration, and, following the Supreme Court's effort to overturn the South's longstanding system of segregated schooling with the *Brown* decision, opposition from Southern congressmen over racial issues easily trumped the need for more schools during Eisenhower's term. Invoking the specter of federal control on the heels of the Red Scare was a common tactic employed by hostile legislators to derail federal aid plans. This provided a blanket justification for congressional inaction that appealed to many constituents suspicious of federal intentions.[58]

Without a well-mobilized public constituency calling for a federal role in the public schools, any bid to gain congressional approval for expanding the National Science Foundation's programs in science education would require a clear overriding national interest.[59] The argument Waterman made before the House Appropriations Subcommittee for Independent Offices in February 1955 was apparently not convincing enough. Following the director's description of the manpower crisis and NSF's proposed plan of attack, the committee chair, Representative Albert Thomas from Texas, interjected his concern about federal control of education. "It is not going to be very long before you have a charge thrown in your face," he exclaimed, "that the National Science Foundation is trying to take over in this country . . . and lay down the pattern and control, through a federal bureau in Washington, the education and training of college students and scientists." Waterman, recognizing the dangerous political ground onto which they had wandered, sought immediately to dispel any misunderstandings of the Foundation's role. "Mr. Chairman," he replied, "I hope we make ourselves quite clear. I am sure our friends and colleagues in science and in the universities know that we are not planning to mastermind the educational pattern from Washington." Waterman reiterated that the Foundation's role was merely to facilitate private initiatives in science education. Seeking, perhaps, to exploit the perceived objectivity and political neutrality of science, he noted additionally that it would not be federal bureaucrats deciding the direction of science education. Instead, he explained, "we are trying to determine what it is *the scientists* feel should be done . . . what they think is the right thing to do."[60] The subcommittee's unease over government interference in local issues, however, remained; it recommended no increase in funds for the education program. Further indicating its disregard for the overall mission of the Foundation, the House subcommittee cut NSF's budget request from $20 million to just over $12 million for fiscal 1956.[61]

To counter congressional reticence, the administration and the press worked to alert an apathetic public to the impending crisis. Following the attention generated by the 1953 National Manpower Council report on scientific manpower, a rising tide of articles on the subject began to appear in the nation's newspapers. Beginning in 1955 the *New York Times* index included a separate category for news concerning the "training and supply of scientists, engineers, and allied technologists," which included over 120 entries. As part of this new attention to technology training, *New York Times* education reporter Benjamin Fine devoted a series of pieces to the worsening manpower situation.[62] In alarmist tones he warned in one front-page story that "while the democracies of the world, including the United

States, are looking the other way, the Soviet Union and its satellites are training scientists and engineers at an almost feverish pace." Like others, he singled out the high schools and the life-adjustment curriculum for the lion's share of blame. While the Soviet's are "stressing science, mathematics, chemistry and physics in the secondary school curriculum," he wrote, "the United States is taking an easy-going attitude."[63] Worse yet, Fine noted, was the fact that, despite the clamor in the press, the public seemed little concerned over charges of a soft curriculum in the high schools.[64]

This concern over the softness of American students was shared by many within the Eisenhower administration. At a National Security Council meeting in the fall of 1955, Secretary of Defense Wilson, in response to a question from the vice president regarding the declining course offerings in math and science at the secondary level, replied that "he was inclined to believe that there was too much of the 'pursuit of happiness' by young people in our secondary schools," that students today were "allergic to the hard work required by courses in mathematics and the sciences." Others continued to insist that the "lack of competent teachers" should not be overlooked as a source of the problem. At the conclusion of the discussion, Nixon suggested that the issue be placed on the agenda of the upcoming White House Conference on Education to raise public awareness.[65]

The White House conference, a nationwide affair held in Washington, D.C. in the fall of 1955, is perhaps the most telling event with respect to the public's engagement with the manpower issue. The Eisenhower administration sought to use the conference as a forum to address the national education crisis and ostensibly get grass-roots feedback on the key issues facing American education from the over 1,800 citizen delegates representing all the nation's states and territories. Conference organizers set up roundtable discussion groups and directed each to come to some consensus on six questions spanning topics from curricula to school financing. Pre-selected subcommittees were responsible for guiding group discussions and drafting statements to be included in the final report to the president.[66] From the beginning, critics questioned the objectivity of the administration in its selection of delegates and organization of the conference. Some claimed, without justification, that the whole affair was "stacked against the supporters of federal aid."[67] Others suspected that the conference was anti-labor.[68] Independent of these charges, where the administration's bias was evident was in setting the terms of the discussion over the substance of the school curriculum.

The curriculum discussion group was assembled to answer the question, "What should our schools accomplish?" Conference organizers chose MIT president James Killian, an administration insider increasingly concerned with the failings of public school science education, to head the subcommittee on this topic. In preparation for his role, Killian consulted with Waterman at NSF on the scientific manpower problem and subsequently circulated among various academics and educational officials a document outlining the competing goals of public schooling.[69] The contrasting views mirrored identically the public debate between the professional educators and the academic traditionalists led by Arthur Bestor; indeed, both Bestor and Harry Fuller were among those consulted on the draft document.[70] Position A, as Killian labeled it, argued that since "many pupils have no interest in bookish matters, . . . and are not able to apply abstract ideas," they should instead "learn as much as they can about how to live happy and useful lives" while they are in school. Position B, however, was obviously the one with which Killian was most sympathetic. "The development of the individual's ability to think and to make judgments," it read, "has been subordinated to his 'adjustment' to society, with the weakening of standards of intellectual performance as the result." The solution was for schools to "narrow their aims in order to concentrate on . . . the education of the young in the basic mental skills and the knowledge needed in the modern world."[71] Bestor commented that, as worded, position B was "so adequate and so forceful that I would have no hesitation in accepting it as it stands."[72] The rather extreme characterizations of the two views left little room for any compromise position.

Killian and the Eisenhower administration must have been disheartened, though not completely surprised, by the sentiment expressed at the conference. More concerned with buildings than curriculum, the delegates called for an influx of federal money with no strings attached, reaffirming their disdain for Washington interference in local affairs. As for their thoughts on the aims of schooling, they decided overwhelmingly in favor of a program that would meet the daily life needs of students and against the academic curricula advocated by the various manpower officials—this even after an opening speech by Vice President Nixon that emphasized the poor state of science education and the pressing threat of Soviet supremacy in scientific and technical manpower.[73] Killian, as subcommittee chair, faithfully conveyed the consensus of the group. His report listed the functions beyond intellectual training the schools had taken on over the past half-century, from programs focused on patriotism and good citizenship to those fostering wholesome family life—programs characteristic of the life-adjustment movement. To the question of whether this "broadening of the goals be recognized as legitimate," the answer was unequivocal: "This Committee answers *Yes.*" "Nothing was more evident at the White House Conference on Education," Killian wrote, "than the fact that these goals, representing as they do an enormously wide range of purposes, are the answer to a genuine public demand."[74]

The very structure of the conference, requiring as it did consensus answers to the questions posed, clearly worked against hopes of formulating any coherent and substantive advice for the president, particularly advice that stressed the importance of the intellect. Chemist and former NSF divisional committee member Joel Hildebrand was frustrated by the effect that the divergent interests of those in attendance had on his group's efforts to lay greater emphasis on the intellectual disciplines in the final report. He likened the process to "trying to deliver a bottle of good California burgundy to a friend in Edinburgh by pouring it into San Francisco Bay, and letting him dip such of it as he could from the Firth of Forth after it had had time to diffuse."[75] In a follow-up report on the White House Conference to members of the National Science Board, NSF director Waterman regretfully noted that "the place of science in the schools of the nation was discussed in a peripheral way only." He added, more hopefully, that only a "very small minority was opposed to Federal aid . . . in any form," implying, perhaps, that the further expansion of NSF education programs was not politically out of the question.[76]

Although the drive for rigorous intellectual training found little support among the general public, the administration had more luck on Capitol Hill, where congressional indifference to science education had dissipated rather quickly over the previous year. Much of this change of heart can be attributed to the publication of the NSF-sponsored book *Soviet Professional Manpower* by Nicholas DeWitt in the summer of 1955. This work provided the first complete, well-documented picture of the Soviet educational system. For many in Congress it was an eye-opening presentation of the specific threat the Russians presented. Earlier congressional objections to expanding NSF education programs centered foremost, as we have seen, on the issue of federal control, but there was also the sense among the legislators that the manpower experts were exaggerating the threat of Soviet proficiency in technical education. DeWitt's book showed that this was no exaggeration.[77]

It was well known that the primary function of the Soviet schools was indoctrination in communist ideology. And just as ideology interfered with objective inquiry and sound scholarship in the universities, many believed that the techniques of indoctrination, combined with ideologically distorted subject matter in the lower schools, provided a kind of intellectual training the United States had little reason to fear. However, as DeWitt argued, with the Soviet march toward industrialization and need for advanced military technology to compete with the United States, the bottom line for their leaders was securing proficiently trained manpower "regardless of whether individual members have or have not been thoroughly indoctrinated." Moreover, in the subjects that mattered most for technological success—the hard sciences of mathematics, physics, and chemistry—"the free application of ideological bias is limited." Thus, DeWitt stressed, contrary to conventional wisdom, Soviet training in these subjects was quite effective. "Laws and the solutions to problems," he explained, "remain the same whether the problem is stated in verbalisms glorifying the Soviet state . . . or contemptuously in words depicting capitalist conditions." While political interference undoubtedly produced some "residual effects upon the mental outlook of individuals brought up under Soviet rule," he concluded, such interference does not affect learning "in fields important for technical development and scientific activity." The Soviet Union clearly was directing her efforts in these very areas of education crucial to national security. Casting an eye to U.S. educational policy, DeWitt noted that "the Soviet educational system offers much more than our own in the teaching of science at the secondary-school level." In fact, he described the "high concentration on science instruction" as the most distinctive feature of Soviet secondary education.[78]

DeWitt's book, along with word that the Soviets had successfully detonated a deliverable hydrogen bomb in November 1955, effected a profound shift in congressional attitudes toward the education programs of the National Science Foundation.[79] Earlier concerns over federal control of education seemed to evaporate in the face of this now real Soviet challenge. At the House subcommittee hearings on NSF funding held in January 1956, Chairman Thomas, no longer concerned with NSF masterminding education, publicly acknowledged his conversion: "This little book, *Soviet Professional Manpower*, I read word for word . . . and after reading it I completely reversed my thinking." The impetus this text provided for congressional support of education reform is evident in his comments. Russia's push in secondary science education, he exclaimed, "is the most alarming situation that I can imagine." "I think I speak for the committee," he continued, "that if we are going to be with the Foundation when the Foundation starts out on some new programs, we are going to encourage and help you."[80] In order to jump-start the nation's schools, Thomas suggested that Waterman print up fifteen or twenty thousand copies of DeWitt's book and "send them out to every high school superintendent in the United States." It would be the "best money you ever spent," he insisted. His concern over the situation was palpable. "This Russian progress is the most startling thing I know of," he repeated. "Lord help us if they ever reach the point where they are ahead of us."[81] After the dust from the hearings had settled, the Foundation's budget for non-fellowship education programs had increased nearly eightfold over the previous year's appropriations to $10.9 million.[82]

The dramatic increase in support for these programs cannot easily be attributed to any single factor. Clearly the primary motivation for Congress to encourage some federal direction in pre-college science education came from the competitive threat of the Soviet technical training programs—especially at the secondary level. The lengths to which the Soviets were willing to go in their quest to achieve scientific and military parity with the United States were alarming to many, as Congressman Thomas volubly noted. This, combined with the conventional view of the Soviet Union as driven by a political ideology bent on worldwide domination, demanded some counter response from the United States in the name of national security.

Not unimportant in the calculus of congressional approval was the ability of the NSF education programs to serve as a conduit for channeling aid to public education, thus sidestepping the contentious issues inevitably raised in debates over general federal aid legislation. Some years later Harry Kelly recalled that during the House appropriation hearings one congressman had asked him off the record whether "Catholic nuns" and "Negroes" were allowed to participate in the teacher-training programs. Following Kelly's affirmative response, he shot back, "How in the hell do you get away with this?" True to the vision of science as politically neutral he replied, "Our job is not any social change. Our job is to increase the scientific strength of the United States"—an answer that, after some thought, seemed to alleviate the legislator's concerns.[83] A relatively uncontroversial path was thus cleared for the simultaneous solution of two problems. Channeling money to NSF science education programs would both work to counter the Soviet technological threat and help reduce, albeit indirectly, some of the financial strain on local school districts, if not materially, at least symbolically. In later years more substantial funds would flow down that initial narrow path.

Although difficult to pin down precisely, it is probably safe to say that the image of science as the objective pursuit of knowledge, immune to ideological corruption, set forth by the scientific community of the time played some role in the relenting of congressional opposition to federal education programs. To counter fears of bureaucratic control of schools, Waterman repeatedly stressed the fact that all NSF programs were guided solely by the expertise and needs of scientists in the field.[84] As Kelly noted, social or political considerations had no place in determining the best course for the advancement of science. Science—especially the hard sciences—possessed a remarkable intellectual integrity, even in the ideological crucible that was the Soviet school system, according to DeWitt's provocative analysis. In the mid-1950s it became possible to argue, as CIA director Allen Dulles did, that science education even had the power in disseminating the rational method of scientific inquiry to expose the false ideology of communism. Dulles claimed that the dual Soviet goals of scientific advancement and thought control would inevitably collide in all disciplines as they had in the field of genetics.[85] Though the technological danger posed by the Soviets clearly presented the most pressing justification for federal involvement in science education, the ideological alignment of science with democracy, in polar opposition to communism, likely provided added reassurance to those who feared subversion in the schools.

## The Foundation Takes Control

By the spring of 1956, the Eisenhower administration had successfully placed the scientific training issue squarely in the public eye. Indeed, it had become a central concern to Eisenhower and his national security advisors as they sought to maintain U.S. technological superiority over the Soviets. While those outside Washington may have been unswayed by the overall import of the situation, members of Congress—at least those holding key committee assignments—were now clearly on board, ready to redress the science education gap that had opened up, particularly in the nation's high schools. Although the National Science Foundation had only reluctantly entered the picture, a recognized need for its expertise and, more important, congressional encouragement provided the confidence necessary for the Foundation to take on the problems of science education and begin to implement its vision for science education reform.[86]

As NSF officials began to think programmatically about how they might improve pre-college science teaching on a broad scale, one of their primary concerns was ensuring that any such programs be in the best interests of the scientific research community. To this end they set out to secure the field for themselves. Perhaps the most immediate threat to the Foundation's plans was the ODM/SICTSE-inspired National Committee for the Development of Scientists and Engineers (NCDSE). The decision to assemble the National Committee—a broadly representative, nongovernmental citizen's group—indicated an understanding on the part of the administration that changes in the nation's educational system, unlike other components of the national defense infrastructure, would require a good deal of cooperation from those with longstanding interests in the schools. What remained unclear was the specific authority such a committee might possess and what particular agenda it would pursue.

A draft of the president's statement announcing the formation of the citizen's committee raised familiar concerns within the inner circle at NSF. It painted the nation's technical training crisis in the alarming colors of impending international conflict. "Either we maintain our present technological superiority and with energy advance our command of the sciences," the draft read, "or we may find the sciences used for our extinction."[87] Foundation officials strongly objected to the tone, sensing that it would incite public demand for the NCDSE to pursue crash programs in scientific training. Though such programs might eliminate the sheer numerical superiority the Soviets possessed, they would accomplish this, NSF believed, at the expense of quality and to the detriment of the long-term health of the American scientific community.[88] As Harry Kelly later remarked, "we cannot produce creative people as we manufacture steel."[89] Ever since the formation of SICTSE, the Foundation had struggled to make its vision of science education the basis for administrative policy in this area. At its core this vision consisted of a gradual, long-range approach to education that emphasized quality over quantity and, in the continuing spirit of Vannevar Bush's *Endless Frontier*, basic research skills over mere technological proficiency. Guiding the NCDSE toward these educational commitments and away from those of technological warfare was a delicate task. Fortunately for NSF, it was well positioned to accomplish just that.

From the beginning, ODM and the Eisenhower administration saw the Foundation as a key resource for the newly-assembled National Committee. Waterman was informed early on that NSF would be asked to provide it with staff services as needed—a relationship NSF quickly sought to bend to its own advantage.[90] At a meeting of the Divisional Committee on Scientific Personnel and Education devoted to discussing the prospect of the new organization, one Foundation official suggested that since the Foundation was not "in a position to recommend against [its] formation," it must instead "be astute and clever enough to integrate its activities with those of the Committee." In working closely with the citizen group it should be possible, he explained, to have Foundation ideas "reflected in the thinking of the Committee."[91]

Influencing NCDSE's agenda proved to be easier, perhaps, than they expected. Work along these lines began with Harry Kelly traveling to Columbus, Ohio to visit Howard Bevis, the president of Ohio State University and prospective chairman of the citizen committee. Kelly discussed with Bevis the expected relationship between NSF and the Committee as well as their respective roles in initiating programs in science education. Essentially NSF would initiate and direct programs and NCDSE would generate public support. As Kelly reported back to Waterman, "my personal reaction to Dr.

Bevis was very favorable. He was not, as we feared, a pompous president of a big college." More important for NSF, he noted, the future chairman appeared to be "very cooperative," an individual "with whom to work." To help Bevis get the proper start, Kelly took the liberty of leaving "some homework for him as background on the problem and on the Foundation."[92] A little over a month later Waterman was invited by the NCDSE executive secretary Robert Clark to provide the initial briefing to the group at their first meeting. In the invitation, Clark encouraged Waterman to "give the Committee as much substantive information as you think they can usefully absorb," since the talk would be the "only one of this sort at the meeting" and would serve as the basis for all subsequent discussion.[93] With NSF providing informational resources, and Waterman framing both the problem and the most desirable solution, by the time the Committee got down to business they had largely assimilated the Foundation's perspective on science education.[94]

The ease with which the National Science Foundation cleared the field of competitors can be linked to the emerging authority of scientists at the national policy-making level. As the Cold War began to turn on long-range issues of technological advancement and scientific sophistication, the expertise of scientific researchers, already held in high esteem in the military-industrial establishment, gained even greater luster in the eyes of Congress. Science education, no longer a trivial matter, required expert attention. Who better to provide that attention than the scientists themselves? The enduring conflict with Russia had drawn education into the high-stakes game of technological one-upmanship. A personal communication from CIA Director Allen Dulles to the National Committee offering any assistance it might need to help ensure "U.S. technological superiority" over the Soviet Union is one indication of the seriousness with which the situation was then viewed.[95] Given this state of affairs it is not surprising that the various manpower committees would defer to what NSF thought best. That is certainly what the Foundation expected. Prior to the formation of the National Committee, one NSF scientist commented that the manpower problem was simply "more complex than a lay committee is in a position to handle effectively."[96] Bevis, the soon-to-be chairman, admitted as much, informing Kelly during their meeting in Ohio that he frankly "did not know much about the problem."[97] Thus, once formed, the citizen's committee was happy to lean on the Foundation all it could for guidance and support.

Over the course of the NCDSE's existence, NSF officials continued to work behind the scenes to ensure that their agenda was not compromised—particularly by anyone in the field of education. Although they publicly spoke of "encouraging understanding between educators and scientists," the scientists felt they already understood the educators all too well.[98] Not long after the announcement of the formation of the National Committee, an official from the National Association for Research in Science Teaching (NARST) contacted the Committee's executive secretary, Clark, lobbying hard for the inclusion of his organization in the mix. There should be "closer cooperation between scientists and science educators," he argued, in working "to solve the shortage of scientists and science teachers." Clark, unfamiliar with the organization, ran the request by NSF staff members for guidance before taking further action. No ringing endorsement was forthcoming. One Foundation official counseled that the "group is OK, but not outstanding." He suggested that "they *might* be helpful, on task forces occasionally." At the very least, he indicated, they should "be given to feel they may be called in for help." Another official commented pointedly that "NARST is definitely more *education*—than science—oriented. . . . Care should be taken to prevent their dominating any of the Committee projects."[99] The message to Clark was unambiguous—educators were to be tolerated only to provide the appearance of cooperation. Any substantive decisions would be made by those within the NSF-led scientific community. Ultimately the National Committee, later renamed the President's Committee for Scientists and Engineers, made no real decisions, which suited NSF just fine. Committee members served effectively during their time drumming up public support for national science education reform, while the Foundation proceeded with its work.[100]

The substantial funding increase Congress bestowed upon the fledgling NSF education division enabled it to mount a systematic attack on the glaring educational problems at the high school level. The division immediately increased the number of summer institutes for secondary science teachers from 18 in 1956 to 91 the following summer, thus bringing updated subject-matter training to teachers across the country.[101] Foundation officials, however, had long recognized that enhanced teacher training would not be sufficient to bring American education into competitive balance with the Soviet system. As Kelly noted, "Clearly the teacher is a key . . . , but the teacher must have the tools, the course content, the curriculum."[102] To this end the division established a program to encourage the development of supplementary teaching aids, which were to include primarily the use of film and television programs in science instruction. In the same spirit, the Foundation sought to fund efforts directed at curricular improvement.[103] Though some had earlier balked at entangling the Foundation in the controversial area of school curricula, with the support of Congress and now ample resources, Kelly began to search in earnest for a place to spend Foundation money. This was the next big project upon which the NSF directors were about to embark. All they needed was a suitable group of scientists who would be willing to develop science curricula along the lines they had in mind.

# CHAPTER 4

# WARTIME TECHNIQUES FOR COLD WAR EDUCATION

The Cosmos Club, located just off of Q Street on Massachusetts Avenue in Washington, was the home away from home for the nation's scientific elite. Informal business was frequently conducted in the parlors of the club's mansion over cocktails, and it was there in July 1956 that Jerrold Zacharias pitched his physics curriculum project to Harry Kelly. Zacharias had initially laid out his plans to develop 90-some instructional movies in physics for high school use in a memo to MIT president James Killian earlier that spring—a proposal that languished even after Killian publicized it in a *Life* Magazine feature article on MIT and the scientific manpower shortage. It was only after Zacharias had shown the memo to NSF director Alan Waterman during a visit to Washington that things began to happen. That same evening Waterman sent Kelly over to the club to work out the details of the proposal.[1] "As we talked," Kelly noted, "most of the members of the Science Advisory Committee joined us. They all were enthusiastic and helpful in convincing Zach that he should take some time off to tackle the problem of a film course in physics."[2] Kelly, perhaps, was the most enthusiastic of all. As Zacharias recalled, Kelly insisted that he undertake the project. "You've got to use up all the money we've got," he said.[3] Before the night was over, Zacharias walked away with assurances of $200,000 to $300,000 of NSF start-up money for his project.

Kelly and the Foundation had been looking for someone to take the lead in science curriculum reform. Frustrated by existing approaches to science education fashioned by educationists, they sought a first-rate scientist to take on such a task. They preferred someone in the physical sciences—someone very much like themselves, who shared the interests of the hard-science elite that dominated the NSF hierarchy. It was an insider's game to be sure, and Zacharias—an MIT physicist and himself a member of ODM's Science Advisory Committee—was one of the key players. Kelly, of course, wasted little time in seizing the opportunity Zacharias presented. Here was an individual with a proven track record, someone NSF could count to bring education into the scientific age. And Zacharias did just that. His approach to curriculum development utilized all the research and development techniques scientists had developed during World War II—techniques that were further honed in the extensive consultation work they did for the nation's defense establishment during the Cold War. Zacharias's goal was to bring the postwar research culture with which he was so intimately familiar to bear on the pressing educational problems of the day. It was at MIT in the fall of 1956 where wartime research models and school science first met in the daily operations of the Physical Science Study Committee (PSSC).

PSSC, however, was just the beginning. The Soviet launch of *Sputnik* in October 1957 saw to that. The Russian satellite landed a hard blow to the American psyche and presented an unprecedented technological threat to a stunned nation. The years of criticism and crisis surrounding the public schools now proved too difficult for an anxious public (at least as represented by Congress) to overlook, and education quickly became the scapegoat for the country's apparent scientific inferiority. Scientist-led curriculum reform efforts were subsequently pushed into the national spotlight as the best hope for recovering U.S. educational preeminence. Under NSF direction, a raft of new projects emerged from the nation's academic enclaves. The first and largest of these post-*Sputnik* projects was the Biological Sciences Curriculum Study (BSCS).

*Sputnik* was a catalyst for numerous changes in American society in the late 1950s. In the field of education, it provided the symbolic threat that led to a wholesale public embrace of the scientific techniques and models that were only just being piloted by Zacharias and his colleagues. In this sense, the Soviet launch served to legitimize the scientists' approach to reform. Indeed, one could argue that the bold innovations of PSSC paired with the international crises of the late 1950s brought about one of the most radical changes in our view of teaching and learning this country has ever seen. All the new curriculum projects—in subjects as diverse as English and social studies in addition to physics and biology—followed PSSC's lead. This was clearly evident in the work of BSCS, which, despite the growing professional tension between biologists and physicists in the postwar period, utilized PSSC as a template for its own project. The case of BSCS illustrates the manner in which the PSSC model was adopted in a discipline far different than physics and by individuals in a scientific profession at the margins of political power during the Cold War. But, more than anything, it reveals just how deeply embedded the physicists' model had become in the government-backed effort to remake science education.[4]

## The Rad Lab Model

Few things speak as powerfully as success, and few human endeavors were as successful—spectacularly so—as the large-scale research and development projects scientists directed during the war. These projects, which emerged from the unique laboratory conditions provided by worldwide armed conflict, initiated an entirely novel way of conducting scientific research, of learning about and interacting with the physical processes of nature. Identifying the essence of these new models of scientific research has been problematic for historians of twentieth-century science. "Big Science" has been the label of choice. But, as Bruce Hevly has aptly noted, "the phrase is conveniently murky, appropriate for an activity that few can define or describe precisely but many feel able to recognize on sight." One might say that whereas science in the prewar years could best be described as small-scale research, pursued with private funds by solitary investigators in the pursuit of knowledge for its own sake, the new science was increasingly driven by interdisciplinary teams of researchers that marshaled both science and engineering expertise in the service of goal-directed projects of national social and political significance. Such large-scale projects depended on high levels of government funding, which was funneled through government contracts to scientists at leading universities in order to preserve their intellectual autonomy and academic working environment. Institutional manifestations of this trend toward large-scale research facilities included the Radiation Laboratory at Berkeley (site of the first synchrotron, or particle accelerator), along with the postwar accelerator facilities at Brookhaven and Stanford.[5]

This new type of scientific work, however, was not defined solely by technical accomplishment (such as that demonstrated by the atomic bomb) or by mere physical size (as in the case of the accelerators that sprang up across the country). One of the most important factors in the success of the research teams during and after the war is to be found in their broad-based, analytical approach to solving the complex problems encountered in modern warfare—problems that involved not just the satisfactory performance of a given piece of technical apparatus, but also consideration of the most efficient use of that apparatus within the larger context of existing weapons technology and of the combatants themselves. This problem-solving approach, which depended on the efficient integration of emerging technologies into goal-directed systems, along with the new organizational arrangements, had a significant impact on subsequent reforms in science education. Of all the wartime projects in which scientists were involved, the one that epitomized this new research model was the Radiation Laboratory at MIT. Examining the details of the organization and operation of the Rad Lab, as it was called, provides important insights into the subsequent forays these scientists made into the field of education.

Scientists began their work at the lab in 1940 under the directorship of physicist Lee DuBridge. Their task was to develop effective systems of enemy surveillance using microwave radar, one of the publicly hailed technological breakthroughs of the war. Institutionally, the lab was housed on the MIT campus and supported through a defense contract with the government's Office of Scientific Research and Development (OSRD). The contract arrangement enlisted the talents of the scientists for the war, while allowing them considerable freedom from military regimentation and control—a freedom of which they took full advantage.[6] From the beginning, the physicists at the lab loosely organized themselves into various research divisions, each responsible for a particular aspect of design or development. The divisions were not rigidly set and allowed for what the scientists saw as essential cross-fertilization of ideas. General tasks to be accomplished were identified in the weekly meetings of the steering committee, which were then undertaken at the discretion of the various research groups. The lack of a rigid hierarchy and the plethora of intellectual talent assembled produced a frenetic, heady, and ultimately productive research environment—an environment that, with the moral imperatives of the war pressing down, fostered a camaraderie that was fondly recalled by nearly all the participants.[7]

The physicists and technicians at the lab covered all phases of research, development, and prototype production. Theoretical models were tested daily in the laboratory and lab results contributed to the revision of theory. Once the microwave devices were perfected in the sterile confines of the lab, physicists had the formidable task of converting them into rugged, field-ready combat tools, capable of being operated by military personnel with only limited scientific training.[8] The vicissitudes of field conditions, especially with respect to temperature and moisture, made such conversions extremely challenging. With help from industrial research groups like Bell Laboratories, physicists quickly developed engineering skills as they were forced to fabricate radar systems that would fit the harsh circumstances of their use. The unique environment demanded a fusion of skills as theorists worked side by side with experimentalists and engineers—each offering insights and correctives to accomplish the technological objective that lay before them.

By 1942, the Rad Lab physicists and engineers had perfected a plane-mounted microwave radar surveillance system capable of detecting enemy submarines at sea. But, despite the impressive ability of this technology to pick out the extremely low profile of individual submarines at the ocean surface, it could do little by itself to win the battle for the North Atlantic. Scientists quickly recognized that how the technology was used in combat was crucial to its ultimate effectiveness, and thus pushed for greater input into the military decisions regarding deployment. This insistence on a greater tactical role bore remarkable fruit in the offensive use of radar to search for and destroy German submarines. The navy initially resisted the overtures of the scientists, contending that the traditional, destroyer-escorted convoy was the best way to protect Allied shipping. Scientists, however, were able to demonstrate mathematically that the probability of pinpointing German submarine locations was greater using radar-equipped aircraft than with any other tactical method. Hunter-killer groups outfitted with radar and following the prescriptions of the scientists dramatically increased successful strikes on German U-boats, thus securing the Allied shipping lanes.

The use of mathematical tools of probability and statistics to improve the efficiency of weapon systems in combat was characteristic of the new field of operations research (OR), and was the seed of the new analytical techniques that emerged from the war.[9] Developed by the British in 1937, OR concerned itself with optimizing weapon effectiveness by altering the variables of its deployment based on quantitative studies of past performance. The OR approach was quickly adopted by American scientists, as in the case of the navy antisubmarine group, and, with contributions from engineers at Bell Labs, was further refined to a point where such analyses provided a clear military advantage to those willing to heed such advice. Following the navy's successes, the other services rushed to establish their own OR advisory groups. "By V-J day, civilian scientists were in vogue as strategic and operational advisors to a degree without precedent in the annals of American military history," observed historian Daniel Kevles. The physicists had become particularly adept at using their technical and analytical expertise to optimize the performance of any manner of complex system to meet carefully defined military objectives.[10]

It was the intersection of all these elements of the wartime laboratories—the loose, free-wheeling organization; the collaboration among top scientists; and the analytical, systematic approach to the tasks before them—that was responsible for the power of the new research model to produce tangible results—results, as we have seen, on which scientists built their reputations as individuals who could get things done. It is not surprising that as they began to turn their attention to the problems of education that they brought with them the organizational resources and analytical skills that had served them so well during the war. This was certainly true in science more generally, where, as Peter Galison has demonstrated, the "war laboratories . . . clearly provided the managerial models, the technical expertise, and even the personnel for the establishment of postwar collaborative laboratory work."[11] It was obvious to NSF official Paul Klopsteg even at the time that "the pattern for military research had become the pattern for basic research."[12] Looking back at the seminal military R&D projects such as the Rad Lab, one can trace a surprisingly direct lineage to the science education reform projects of the 1950s and 1960s.

## Enduring Organizational Patterns

The transition from these defense projects to the Physical Science Study Committee in organization and personnel was in many ways seamless. The summer study was key to that transition. A variety of government agencies charged with maintaining U.S. national security came to depend on the expertise scientists had developed, and the summer study was one of the more interesting conduits for channeling that expertise to government officials. It was an ad hoc consultation arrangement developed at MIT, that in many ways sought to recreate the heady intellectual environments of the wartime weapons laboratories. The most significant of these was Project Hartwell, completed at MIT in 1950 with funding from the Office of Naval Research (ONR). The Hartwell scientists were assembled under the direction of Zacharias to examine the threat of Soviet submarine attack. Their report so impressed navy admirals that thereafter they referred to it as the "bible of underseas warfare."[13] As two ONR officials later wrote, "In matters of approach, conduct, and impact, it was to become exemplary for the studies of the next ten years."[14] Its influence on Zacharias was no less profound. He reportedly claimed that "every planning activity he ever ran after 1950 was modeled after Hartwell."[15] PSSC was no exception.

The military, however, was not the only group seeking the expertise of the scientists in the early 1950s. Only days after the navy pulled out of MIT at the conclusion of Hartwell, the State Department moved in. Undersecretary of state James Webb, well aware of the services MIT had been providing to the defense department, had earlier approached Institute president James Killian about the possibility of establishing a study to examine the problem of Soviet jamming of Voice of America radio transmissions.

These discussions led to the organization of Project Troy—named for the ancient city conquered with the aid of a wooden horse. This project is particularly useful in giving us insight into both the defining characteristics of these summer studies as well as the government's willingness to engage problems that extended beyond narrow technical and military concerns to areas that bordered more closely on problems related to education.

Troy was part of a comprehensive effort of the Truman administration to counter the widespread propaganda offensive then being waged by the Soviet Union. The Cold War at the turn of the decade, as we have seen, was primarily a battle of competing political ideologies, of antithetical social and economic visions. What was needed, in Truman's estimation, was a great "Campaign of Truth" that would make the United States and all that it stood for known to the world as it really was, not as it was depicted by the enemy. The charge to Project Troy was to find the most effective means for disseminating the American point of view to the people of the Soviet-bloc countries. In early November 1950, the study participants settled into the same Lexington facility that housed Project Hartwell only a few months earlier and began to consider the most effective ways of achieving this goal.[16] In the end, the scientists proposed to the State Department a coordinated effort that included, in addition to radio transmissions, the use of balloons, motion pictures, student exchanges, library services, and even export commodities such as pharmaceuticals, flashlights, and fountain pens as a means of spreading U.S. propaganda.[17]

The organization and daily operation of Project Troy exemplified the new collaborative approach propagated by the scientists. It was, in fact, touted by the study organizers as the most effective way of solving the complex technical and political problems that came to define the Cold War struggle with the Soviets. "In essence," the report stated, "this involves the establishment of a manageable, full-time, ad hoc group relieved of their duties so that they may for three months or so give their undivided attention to the problems involved." Furthermore, it went on, "it involves placing them in a building where they can work under appropriate conditions of security free from conventional interruptions."[18] Participants were encouraged to take advantage of accommodations provided at a single site in order to maximize the number of informal contacts and conversations, which were enriched by the variety of disciplines represented.[19] Perhaps the foremost criterion for success, though, was that the participants themselves be "the most capable and well informed American scientists and specialists that can be found."[20] Zacharias, who was a consultant to the project, always insisted that it was the people who were too busy to commit to a summer study that you really needed if you wanted it to succeed.[21]

When Zacharias decided to take on high school physics education he did it the best way he knew how—he organized a summer study. PSSC thus fell in line with Hartwell and Troy before it as a study of the highest national priority requiring the sustained intellectual attention of the nation's scientists. For Zacharias, who had professionally come of age at the Rad Lab and was later heavily involved in MIT's various postwar defense laboratories, no other model would do. Not long after his meeting with Kelly at the Cosmos Club, he began gathering the appropriate personnel to tackle the problem. Zacharias immediately tapped the available local talent: MIT physics department chair Nathaniel Frank, MIT president James Killian, war research patriarch Vannevar Bush, and from the Manhattan Project, Philip Morrison and Francis Friedman. The group was rounded out by Polaroid founder Edwin Land, Educational Testing Service president Henry Chauncey, and Nobel prize-winning physicists and Rad Lab alums Edward Purcell and I. I. Rabi. All, of course, were well versed in defense-related research and development projects. Zacharias and Friedman, the individuals who made up the heart and soul of the project, had worked together on both Hartwell and Troy, as did Purcell. The lessons they learned from those experiences, particularly what Friedman had taken away from Project Troy with respect to information dissemination, would serve them well in their new educational endeavor. The connections to Troy would draw even tighter in 1958 when James Webb, the State Department official who had initiated Project Troy on behalf of the Truman administration, took over the administrative reins of PSSC upon its conversion to Educational Services Incorporated.[22]

The eventual steering committee membership was a virtual who's who of the American scientific elite.[23] For both Zacharias and NSF, this was key. Zacharias was confident that nearly any problem could be resolved if, as he said, "first-class intellect" was brought in to do the job.[24] This was the lesson he had learned from his extensive work as a government consultant. As for the Foundation, it had long been fighting against a dilution of quality, opting at every turn for policies that would support the top scientists of the country to ensure that "best" science always won out over "more" science.[25] This commitment applied to science education as well. During their meeting at the Cosmos Club, Kelly implored Zacharias to take up the challenge of curriculum reform lest the project be "invaded by second- and third-raters."[26]

Even before the details of the project had been finalized, members of NSF's Divisional Committee for Scientific Personnel and Education granted Zacharias's group "their wholehearted approval."[27] Such approbation was directed as much at the method of attack as it was at the specifics of the curriculum plan itself. Pointing to the important work he had done previously in the name of national security, MIT chancellor Julius Stratton insisted to Foundation officials that Zacharias could do for education what he had done for the defense department.[28] In the cover letter that went out with the grant proposal to NSF that August, Stratton wrote: "Professor Zacharias has a magnificent record of success in conceiving and carrying through such intensive study projects to fulfillment and we shall count again on his leadership."[29] Subsequent funding pitches often traded on the shared experience and past success with defense problems as justification for support. In briefing members of the National Academy of Sciences (NAS) on the course in 1958, Killian recounted how Zacharias "had led a group of study projects concerned with defense problems," and that what had emerged from these was a technique whereby "creative, imaginative people, free from interruption, [were] encouraged to think in an uninhibited way." "Here was the opportunity," he explained, "to try to provide [this] set up and approach to the problem of teaching in secondary schools."[30]

PSSC embodied many of the attributes of the large-scale research projects that sprang up during and after the war. The physics curriculum project was certainly big in terms of funding. The initial grant of $300,000 was just over the minimum amount of money ($250,000) Zacharias felt was required to do any project well—a threshold of funding that some at MIT teasingly referred to as a "Zach."[31] PSSC cost the government a total of $5.7 million by the spring of 1964. The Course Content Improvement Program at NSF ultimately spent upwards of $18 million on such projects across the sciences during this same time period, an amount that, although small by science standards of the day, was unprecedented in education.[32] (Though even that, Zacharias felt, was too low.[33]) As with the other federally-funded projects, Zacharias had also clearly tapped into an area of keen interest to the scientific community, Congress, and the national security agencies of the Eisenhower administration. All parties concerned were eager to move forward in the long stagnant area of pre-college science curriculum.

The National Science Foundation made its usual funding arrangements, setting up a government contract for the first phase of the curriculum study.[34] The arrangement was one to which MIT faculty had grown accustomed and the scientists preferred. In a letter soliciting additional funds from various private foundations, Zacharias highlighted the desirability of these particular institutional arrangements. "Many of the members of the Physical Science Study Committee," he wrote, "are already familiar with patterns (largely established during World War II and continued thereafter) which permit a harmonious relation between the companies which execute the ideas and the individuals or groups who, working under public funds, conceived them in the first place."[35] The harmonious relationship to which Zacharias referred was one in which scientists received generous funding and then were allowed free rein to work as they saw fit. For NSF, the contract was the essential legal buffer that insulated the Foundation from potential congressional charges of "masterminding" the pattern of education in public schools. It allowed them to claim, as Waterman had argued before Congress, that they were merely helping scientists do "what it is *the scientists* feel should be done."[36]

Informally, the business of funding and organizing the physics curriculum project occurred as did most research after the war, through the extensive network of personal contacts that made up the infrastructure of the scientific community. Nearly all the PSSC participants and NSF administrators could trace their lineage back to the Rad Lab or Los Alamos. Harry Kelly knew Zacharias from their radar work during the war. Stratton was himself a member of the National Science Board, the overseeing body of NSF, and Killian, Land, Purcell, Rabi, and Zacharias were all members of the ODM's Science Advisory Committee along with Zacharias's "old friend," NSF Director Waterman.[37] This shared social and professional environment factored in to the Foundation's support of Zacharias. As Kelly later commented, one of the reasons the PSSC funding went so smoothly was that at his first meeting with Zacharias "some of the board members [National Science Board] were already there—part of the scheme, you know."[38] When the deputy director of NSF expressed concern that the initial PSSC grant had been awarded outside established peer review channels, Waterman replied that for all intents and purposes it had been properly reviewed, counting an informal discussion with the Science Advisory Committee (perhaps the Cosmos Club discussion over cocktails) as one source of external review. NSF's level of comfort with Zacharias stemmed, no doubt, from the fact that he was an equal member of the scientific fraternity. Kelly and the rest of the NSF hierarchy were confident that the physics curriculum project would be in the best interests of science as they saw them. Given this relationship, NSF essentially let Zacharias take the lead in developing the specific pattern for course-content improvement; Foundation policy in this area was then drafted accordingly.[39]

Once the project formally got underway in September 1956, the MIT group concluded that a film project alone would fail to create the kind of intellectual environment it was seeking for the high school classroom.[40] The Steering Committee decided that a more comprehensive physics curriculum package would be necessary to supplement the instructional films proposed. That December, the Committee convened a three-day conference on the MIT campus during which it solicited input on course content from representatives from Bell Laboratories and both the Cornell and Illinois physics departments.[41] What emerged by February the following year, in addition to the films, were plans for a complete textbook covering an entire year of physics, a detailed teacher's manual, a series of ancillary monographs to elaborate on technical details left out of the text, new laboratory exercises, and examinations to assess student learning. The group set September 1958 as the target date for completion of first drafts of the course materials. Zacharias estimated that they would need up to an additional $2.5 million to bring the project to completion.[42]

The expanded plans required some obvious division of labor. Zacharias put Francis Friedman in charge of the textbook writing. Chapters of the planned book were further divided among volunteers from Bell Labs and the various universities.[43] The key addition to this group was Cornell theoretical physicist Philip Morrison, a veteran of Los Alamos, former student of Oppenheimer, and brilliant science writer.[44] In addition to the textbook group, the Steering Committee established a laboratory group headed by MIT theoretical physicist Uri Haber-Schaim to develop innovative lab materials and procedures along with a monograph group under the direction of Laura Fermi, widow of the late Nobel laureate Enrico Fermi. Finally came the film group, headed by Zacharias with the help of Stephen White, a former *New York Herald Tribune* science writer and close friend of I. I. Rabi. Together they constructed a film studio from scratch not far from Cambridge following the termination of their association with Encyclopedia Britannica Films (as a result of creative differences), with whom they had initially contracted.

Communication among the various parts of the project was haphazard at best. All decisions ultimately had to meet the approval of the core working group, which consisted primarily of Zacharias in consultation with Friedman, Morrison, and White.[45] Each group, however, was trusted to take their work in innovative and promising directions as they arose. Meetings were often informal hallway discussions, decisions were made on the fly, constrained only by technical possibility and a vision of the final product.[46] All of this contributed to the project's creativity, which one outside reviewer believed was the "dominant characteristic of PSSC operations." The reviewer further noted, though, that this creativity existed in constant tension with the unstable atmosphere of a "'crash' program." Both contributed, in his estimation, to the fact that "in its methods and its end product, PSSC is a disturbing force to the educational community."[47]

In 1957 Zacharias moved from the old physics building on the MIT campus to the newly constructed Compton Laboratories. From his new office on the fourth floor, he now looked out over Building 20—the hastily built wartime home of the Radiation Lab. This proximity to the old Rad Lab was not merely physical. Successful or not, there could be little doubt that Zacharias had by that fall recreated in PSSC the same broad organizational structure and intellectual atmosphere that had so captivated the radar pioneers during the war.

## Systems Engineering and the Classroom

The participation of world-class scientists in an open, flexible organizational structure was only one part of what made PSSC so innovative in the field of education. More significant in the development of the course materials was the utilization of the systems-engineering perspective. It was this analytical approach that had made the various summer defense studies on which PSSC was modeled so highly regarded. Systems engineering involved the consideration of not just the optimum performance of a given human/technological system, but the entire array of possible alternatives that might be created using existing or newly-developed technologies. Scientists using this approach, in other words, had the freedom to design new systems from scratch without the constraints operations-research analysts faced. The postwar labs, particularly those at MIT, expended a great deal of effort on the development of such systems for the defense establishment. The potential of this approach to meet increasingly complex military and even social needs was explored in a variety of institutional settings beyond MIT as well—from the Air Force think-tank RAND, which among other things exhaustively calculated multiple thermonuclear war scenarios, to the National Research Council's Committee on Operations Research, which called for the application of systems thinking to nonmilitary areas such as traffic management and industrial retooling.[48]

The success of both Hartwell and Troy has been attributed to the effective utilization of this analytical approach. Hartwell had been commissioned initially to study the problem of Soviet submarine detection. Before it began, however, Zacharias insisted that changes be made in the study's objectives. He felt that the Hartwell question needed to be broadened to take advantage of the operations perspective that had been employed so effectively during the war. The problem the navy should be interested in, Zacharias insisted, was not submarine detection, but rather the overall security of overseas transport. The final report contained a range of recommendations related to this broader question of the navy's mission in light of the integrated systems involved.[49] Similarly, the Troy group took it upon itself to advise the State Department not simply about information dissemination techniques, but rather the most effective means of waging what it termed all-out "political warfare," the very concept of which required a systems point of view. An information program by itself was seen as "relatively useless." Such a program, the participants argued, could only become effective when "designed as one component in a political 'weapon system,'" which needed to take into account, in addition to the available technologies of radio transmission and telecommunication, things such as the structure of Russian administrative systems, the broader cultural disposition of the Soviet people, and the psychology of mass persuasion.[50]

Systems thinking had been with Zacharias since his experiences at the Rad Lab and as a member of the Hartwell and Troy summer studies. The seeds of applying this analytical approach to the problems of science education no doubt were sown during the early meetings of the Science Advisory Committee (SAC), where reports of scientific manpower shortages were presented alongside those touting the benefits of the systems approach to defense studies. Both topics seemed to crop up regularly in the early days of SAC. Zacharias recalled that the people from the military "would come in and complain that the Russians were getting ahead of us, that we had to do something about education, about teaching, getting more engineers, more scientists."[51] At a meeting in 1952, in fact, Zacharias himself reported to the group on the work and approach of Project Lincoln, a systems study concerned with air defense, which was followed immediately by a working lunch with ODM-head Arthur Flemming, who discussed the latest problems related to scientific manpower and education.[52]

From the beginning, the physicists on the PSSC Steering Committee sought to take advantage of the systems approach in the field of education. This was the primary reason that the initial film project was expanded to include a textbook, outside readings, a laboratory program, and other materials. Zacharias was convinced that integration of both subject matter and instruction would be central to producing a solid course. "The most effective way of teaching," one early report outlined, "is to use several methods and media concurrently. Some parts of physics, like the evolution of ideas, mathematical deductions, etc., can best be learned if *read* repeatedly. . . . The significance of physical phenomena, on the other hand, will best be understood if the phenomena are *seen* repeatedly. . . . And the experimental method can be mastered both by seeing how demonstrations are prepared and carried out, on film and in the classroom, and by actual experimentation in the laboratory."[53] When asked at a briefing of NSF's Division of Scientific Personnel and Education whether the PSSC approach was more than just the revision of the substance of high school physics, Zacharias replied that "he considered it an experiment in both content and in method," and hoped that physics teaching would be improved with respect to both as a result.[54]

A key characteristic of any such project, in defense systems in particular, was the integration of emerging technologies into the human/machine systems developed.[55] In the PSSC program, the motion picture component provides an excellent example of just such a technology. Instructional films had been used increasingly during World War II for both skill training and political indoctrination of military personnel.[56] By the 1950s, researchers in education and psychology saw film as a powerful means of addressing the shortage of qualified science teachers.[57] Henry Chauncey, the president of Educational Testing Service and soon-to-be member of the PSSC Steering Committee, made the argument that the ideal solution to this problem "would be to have fewer teachers than at present and utilize them to better advantage than we now do." In the matter of imparting knowledge to students, Chauncey claimed that "instructional films can do as good a job in this respect—if not better—than the average classroom teacher." In this way, the very "best teachers in the country" could be reproduced at will, and the celluloid distributed wherever needed. The day-to-day classroom tasks could then be handled by the classroom teacher or even a clerical assistant. The primary pedagogical task, the conveyance of content, in this scheme would be "given over, in effect, to the experts who prepare the instructional films . . . and accessory materials."[58]

In the spirit of Chauncey's vision, Zacharias and the rest of the PSSC group devoted the bulk of their time and money to the production of the film series. To ensure the highest quality product, they consulted numerous film industry representatives including such renowned filmmakers as Frank Capra and Walt Disney.[59] Capra, whose experience with instructional films derived from his work producing the *Why We Fight* series of propaganda films for the Research Branch of the army's morale division during the war, stayed on as a member of the Steering Committee through the publication of the first edition of the textbook.[60] On the technical end of things, the group, at the urging of Polaroid founder and Steering Committee member Edwin Land, seriously considered shooting the films in 3D Technicolor to improve the realism of the presentation, a technique that would have literally placed research physicists in classrooms across the country in all but the flesh.[61]

Though never widely used, the project also developed a "school-proof" projector, which had a variable projection speed that reduced the total amount of footage needed per film, and thus the cost to schools. It was loaded using film cartridges in order to eliminate the need to thread the film through the usual complex of sprockets—this, Zacharias commented, so that even "the least talented football coach can get his physics class under way without a fumble." There were also plans to use a strong enough light source in the machine to permit partial illumination of the classroom "for purposes of a little more note-taking and a little less pigtail tweaking than usual."[62] The ultimate fruit of this filmmaking effort for the physicists was to be direct access to the classroom via a medium that gave them complete control of the pacing and content of the physics taught.

But, as confident as they were in their ability to produce intellectually compelling physics presentations on screen, they worried about what would happen in the classroom when the projectors were turned off. "I am sure," Zacharias commented to Webb, "that if we leave undone the job of converting a twenty-minute film into a fifty-minute period, we are going to get all sorts of funny results."[63] Both Zacharias and Friedman were specifically concerned with classroom discussions being "taken off into side alleys by teachers who did not deeply understand the material that was being taught."[64] They also recognized the difficult position in which the films might place the teachers. Zacharias wanted to be sure that the teachers' authority was not undermined; they at least "have to appear wise" to the students, he insisted.[65] This was where the supporting materials came into play. By adding carefully designed ancillary materials—the textbook, problem sets, demonstration materials, wall charts, and lab experiments—they hoped to enmesh the teacher in a web of technical and textual support that would presumably ensure both teacher and student success according to their dictates. Especially important was the teacher's manual, a monstrous document over 1,000 pages long that provided a day-by-day roadmap for teaching the course; it became, in a sense, the operations manual for the PSSC instructional system.[66]

### The Woods Hole Conference

In September 1959, Zacharias, Friedman, and curriculum reformers in mathematics and biology met with psychologists and other experts at Woods Hole in Massachusetts to take stock of the educational reforms that were taking place. The Woods Hole Conference, formally called the "Study Group on Fundamental Processes in Education," was conceived by the NAS Advisory Board on Education as a summer study no different than those before it. Headed by Harvard psychologist Jerome Bruner it was held over the course of ten days at the Whitney Estate, a secluded facility with a history of hosting defense studies for the Air Force, and funded in part by NSF, the RAND Corporation, and the Air Force Research and Development Command.[67] The initial NAS proposal laid out the by now familiar Zacharian argument that the most effective way to facilitate education reform was to bring "together imaginative people from various disciplines, engage them in provocative discussion, and challenge them to bring all their talents to bear on fundamental questions in education." In the intense environment at Woods Hole, the insights of anthropologists, social scientists, and psychologists would combine with "the scientist-engineer with a wide range of competence in communication theory and electronics . . . to bring a fresh approach to problems of school learning."[68]

Although long-range planning for research in education was prominent on the Woods Hole agenda, there was the sense among participants that, in addition, the conference should serve a "crash program" type of function in its charge to provide immediate support for the ongoing curriculum projects, which by this time included biology and mathematics.[69] Bruner noted in a pre-conference memorandum that "it is worth drawing a parallel . . . to the experience of psychologists working in the armed forces at the beginning of World War II. Both with respect to training devices being used and in the design of instruments to be used in combat and support operations, it turned out to be the case that a great deal of useful work could be done without the support of new research." He went on, "I rather suspect that at the outset there will be a parallel in work on curricula."[70] As a veteran of the Psychological Warfare Division of the Office of War Information, Bruner spoke from experience. He had also gained valuable insights into the dominant research and defense consultation models as one of the 20-some experts who were enlisted to work on Project Troy, as did MIT historian Elting Morison, and, of course, Friedman and Zacharias, all of whom joined Bruner at Woods Hole.[71]

In keeping with the systems studies of the time, the specific objectives of the conference as laid out in the initial proposal were the: "1). Improvement of Communication of Knowledge of Subject Matter, [and] 2). Improvement of Use of Technological Devices and Systems for Education."[72] These goals provided the analytical framework that focused much of the intellectual give and take over the course of the ten-day meeting. The newer instructional technologies were prominent on the agenda. Films from the various projects were screened in the evenings, and two afternoon plenary sessions were devoted to teaching machines and audio-visual aids. All of these various material technologies were brought together within the systems vision in the report of the teaching-learning systems group, which by the end of the conference was renamed the Panel on the Apparatus of Teaching.[73]

The Apparatus group was clear about the promise the systems approach held for improving education. The members lamented "the disparity between the current levels at which the educational enterprise uses modern technology and the levels of its use in commercial communications, in transport and industry, or even in agriculture." Education, they felt, with its "resistant conservatism," was more than a step behind the times in this regard, and that the "adoption and exploitation of a systems approach to educational design would further the application of modern technology to the improvement of education." There was no reason, they argued, not to take advantage of approaches that have proven themselves time and again in other areas. "In our present day society, tremendous forward strides have been made in the design and development of new technical integrations of men and machines in the form of systems." Although education did not fit exactly "the most spectacular examples of system design . . . found in complex military situations," they claimed that schooling at both the elementary and secondary levels did represent "the kind of complex organismic enterprise the improvement of which can aptly be planned according to system development principles."[74]

To help realize this vision, the panel provided a ready-made systems template for curriculum design. In developing any course, the authors explained, one must begin with the definition of the course goals. Once the goals are set, the "functions to be performed by the various components" of the educational system could be determined. These naturally included such things as knowledge transmission, student motivation, attitude and skill development, and achievement diagnosis. Following that, "consideration can be given . . . in detail to the problem of assigning functions to men (generally, teachers) and machines in such a way as to optimize the effectiveness of the whole system."[75]

Comparisons with weapon systems were quite explicit during the conference sessions. In a memo summarizing the presentation of the Apparatus Panel's report, Bruner noted that "we introduced this subject for discussion today by suggesting the analogy to a weapon system—proposing that the teacher, the book, the laboratory, the teaching machine, the film, and the organization of the craft might serve together to form a balanced teaching system."[76] This followed the recommendation of an air force study on education and training media convened by the NAS only one month earlier, the report of which was circulated to the Woods Hole Conference attendees. That study similarly concluded that "the goals of education . . . expressed in terms of the human functions and tasks to be performed . . . can be as exactly and objectively specified as can the human functions and tasks in the Atlas Weapon System."[77] The modern Cold War weapon system was, in the minds of all these reformers, the epitome of rational instrumentation—a powerful model to be emulated in seeking solutions to educational problems.

The strategic use of film technology to address the physics teacher shortage and its integration into the broader systems-engineering perspective are but two of the most characteristic examples of the new postwar research approach that Zacharias and the PSSC Steering Committee brought to the problems of science curriculum development. The arrangements they established both with MIT as their institutional home and the National Science Foundation as their principal funding agency, the continuity in the use of the summer defense study as a means of attacking the problems of high school education, along with the fast-paced, loosely-structured working environment of the project itself, reveal the depth of the connection the group had to the established physics research community in the United States.

Clearly, the well-defined patterns and models of scientific research that were exploited to national advantage for military purposes during the Cold War were brought over and applied with few modifications to the field of education. The striking parallels I have drawn here, interestingly enough, were not lost on the participants themselves. In a tribute to Francis Friedman, who died unexpectedly in 1962, Philip Morrison commented on the new vision PSSC had brought from science to curricular research and development. "We have seen," he described, "a change occur in the research laboratory, and now the same processes are plainly at work in the classroom. This teaching of a new sort is analogous to the new sort of laboratory." The shift in approach, as Morrison saw it, was profound. "Just as the Radiation Laboratory at Berkeley [home of the first synchrotron] has become a kind of symbol of the great laboratories of our time . . . so around the group at MIT, six or seven years ago, a new order of magnitude was entered in teaching. For the synchrotron, read the film studio; for the teams of theorists and experimenters, substitute the many people of diverse backgrounds brought together in the committees, panels, summer studies characteristic of this method."[78] Looking back some years later, Zacharias described his primary contribution in PSSC as bringing a systems-engineering approach to the problems of education. "I did it," he stated flatly—"there's no question."[79] While the goals of postwar scientific research and curriculum design were certainly distinct, the methods the physicists had drawn upon were essentially the same.

## Accommodating the Biologists

The appearance of *Sputnik* in 1957 catalyzed the National Science Foundation's movement beyond physics curriculum reform into biology, chemistry, and numerous other subjects. For the projects in these disciplines, PSSC served as NSF's exemplar. More than that, the public prestige accorded physical scientists and their placement at the head of both the National Science Foundation and the science advisory apparatus of the federal government allowed Zacharias and other key members of the PSSC Steering Committee to directly shape the formation of the subsequent curriculum projects. This influence was particularly apparent in the origination and operation of the Biological Sciences Curriculum Study, which conformed in many respects to the curricular research and development model established at MIT.

The relatively late arrival of BSCS was more a product of NSF indifference than lack of interest on the part of the nation's biologists. On the contrary, biologists had been kicking around ideas for improving biology education as early as 1952, and a concerted effort to reexamine biology education began in the spring of the following year under the auspices of the NAS Division of Biology and Agriculture.[80] With some start-up funds from NSF and a more sizeable grant from the Rockefeller Foundation, the NAS established the Committee on Educational Policies (CEP) in 1954 to begin work in this area.[81] For three years the Committee focused on how best to reform biology education, especially at the high school level. The group's most significant product was a compendious sourcebook of laboratory and field-study material prepared with the help of high school and college biology teachers during the summer of 1957 at Michigan State University.[82] Representatives from NSF monitored the progress of the Committee but offered no more than modest support of their work.

The Soviet satellite provided the necessary spark for more immediate action. In its wake, the CEP put together a detailed proposal to revamp the high school biology curriculum, and the American Institute of Biological Sciences (AIBS) (a national umbrella organization of biology societies with its own education program already underway) moved to collaborate more closely with the NAS committee in its work.[83] To get things moving on the federal end, the CEP issued a statement in March 1958 unanimously recommending that "the National Science Foundation invite a small, carefully selected, group of outstanding biologists . . . to advise the Foundation as to whether there is need for a major program to prepare new instructional materials in secondary-school biology."[84] The Committee obviously believed that such a need existed. The widespread public attention that *Sputnik* had focused on secondary education prompted NSF to act sooner than it might have otherwise. Although comfortable with their venture into high school physics, Harry Kelly felt that "as a result of *Sputnik*" the Foundation was "forced into supporting studies of curriculum" on a broadscale basis.[85]

NSF soon reconciled itself to the fact that it had to sponsor additional projects, and biology was at the top of the list. The Foundation's primary concern, however, was not just to fund any project, but rather to fund only "studies of course content by highly competent groups." The high standard Foundation officials set was naturally desirable in and of itself, but was also necessary, they felt, to avoid any public backlash in this "delicate area for federal support."[86] The question of educational quality was uppermost in Zacharias's mind as well. *Sputnik,* he worried, was simply going "to draw more crap into the vacuum" than anyone could imagine.[87] In light of these concerns, NSF "spoke strongly of the need for major program[s] to improve the content of biology and chemistry teaching, *comparable to the M.I.T. program in physics.*"[88] The problem was finding the appropriate personnel to do the job—the "leading scholars in each field" as one memo put it.[89] In discussing the Course Content Improvement Program at NSF, Kelly often touted the intellectual power of "a group, such as Zacharias, Rabi, Bethe, Bode, Purcell, Rossi and Schlichter," directing the reform of physics education.[90] The biological counterpart to this group was not as clearly evident. A proposal in the works from the National Association of Biology Teachers, principally a secondary-school organization, to develop an outline for a "good high school biology course," in NSF's eyes, did not even come close.[91]

Although aware of the activities of the NAS Committee on Educational Policies, Kelly began his search by soliciting from Zacharias a list of, in his words, "biologists who would be effective in working out a 'Zach' project."[92] Twenty-two names made the cut, including such luminaries as Caltech geneticist George Beadle, Yale limnologist G. Evelyn Hutchinson, and Berkeley geneticist Curt Stern. None apparently was willing to take the necessary time to head up a biology curriculum project similar in magnitude to PSSC. There things stood over the summer. In September 1958, Kelly informed Zacharias that they had one proposal for such a project that they had been aware of since May, from the botanist Hiden Cox, the executive director of AIBS. Before making any decisions, however, Kelly had planned to consult an outside group of biologists to get their input regarding the AIBS's potential for success.[93]

Officials at AIBS were eager to get their project started. On more than one occasion, Cox made inquiries regarding the status of their proposal. The Foundation, though, seemed to be dragging its feet, waiting first to see whether biologists of sufficiently high status would commit to the project before making any formal grants.[94] Five biologists from Zacharias's initial list did finally agree to participate, at least in the early phases of the work. With pressure mounting, NSF took the best of what was available and finalized a grant to AIBS of $143,200 (less than half of PSSC's startup grant) for a pilot study to develop a new high school biology curriculum on the last day of 1958. Zacharias was kept up to date.[95]

The first meeting of the Biological Sciences Curriculum Study convened on February 5, 1959 in Boulder, Colorado—the location chosen primarily for its central location and the agreeable climate it would provide scientists during intensive summer work. The group of biologists AIBS had assembled was far from the third-rate bunch Zacharias and Kelly had feared. Florida zoologist Arnold Grobman was tapped to direct the study. Bentley Glass, Johns Hopkins geneticist and member of the American Association for the Advancement of Science (AAAS) Board of Directors, chaired the Steering Committee. Also in attendance were Wisconsin plant pathologist and AIBS president James Dickson, Rochester physiologist Wallace O. Fenn, Penn botanist David R. Goddard, and Yale ecologist Paul Sears.[96] Outside reviewers generally agreed that this was a "very competent group."[97] As they settled in to work over the next few meetings, there was the inevitable turnover as members came to appreciate the magnitude of the task.[98] After some unsuccessful attempts to restock BSCS with a few more nationally recognized biologists, Glass and Grobman appeared to settle for those scientists who were willing simply to put in the necessary time and effort to do the job well. That should have been our plan all along, Glass insisted, "instead of wasting time with big shots too busy to enter into our program."[99]

Whether attributable to the quality of personnel or just the hard-science bias of the physicists, Zacharias, Kelly, and others at NSF never developed great confidence in the BSCS project. After reviewing and having others review early drafts of the biology text materials, Zacharias openly questioned whether "all of our course curriculum development money was being spent wisely."[100] To be fair, this sentiment extended across the board to nearly all the post-*Sputnik* curriculum projects. Kelly let it be known to PSSC members that NSF had "a real concern about how the mathematics, chemistry and biology groups are proceeding."[101] From his perspective, few of the groups were able to match the "life, sparkle, or inventiveness" of the physicists.[102] Subsequent financial irregularities at BSCS, along with other minor missteps, appeared to confirm many of these suspicions.[103]

Though the biologists failed to inspire the same level of confidence in NSF that the physicists did, it was not for lack of the appropriate approach. They clearly followed, deliberately or not, the pattern of curriculum development PSSC had established. Like PSSC, BSCS had assembled a steering committee of top scientists, such as it was; found an institutional home at the University of Colorado; and began receiving funds from NSF (well over $6 million by 1963) through AIBS, the official organizational sponsor of the study.[104] Many of these arrangements were, of course, dictated by the policy requirements of NSF, which had adopted rather strict funding guidelines to insure its political insulation with respect to public education. But BSCS internalized the physicists' perspective well beyond the administrative structure required by the Foundation, and it is not difficult to see why this occurred. Glass was certainly familiar with the more recent operational methods of the scientific elite from work he had done with the high-profile NAS Committee on Radiation Hazards Due to Fallout, as well as other national committees and organizations.[105] BSCS biologists, more generally, were also aware of the kind of goal-oriented research characteristic of the physical sciences the federal government increasingly countenanced.[106]

As the agency coordinating the curriculum development projects, NSF did its part by repeatedly holding up the PSSC work as a model for course-content studies in every subject area.[107] The first BSCS Steering Committee meeting, following general introductions and other administrative matters, opened with representatives from the Foundation providing an overview of the MIT program.[108] And, of course, the entire Woods Hole Conference in which Grobman and Glass had participated was designed to promote the systems approach being implemented by PSSC. Kelly clearly saw the physics project as the keystone to broadscale curriculum reform and never hesitated to share with the other projects his belief that the money spent on PSSC was "one of the best investments this country has made."[109] NSF's heavy-handed administration of the biology project's funding also undoubtedly created strong expectations to conform to the physicists' methodology. NSF representatives attended every meeting of the BSCS Steering Committee as well as many of those of the various working committees, a level of surveillance to which PSSC was not subjected. Throughout the curriculum development process, Foundation officials did not hesitate to let their dissatisfaction with sundry aspects of the project be known, which often came, as Hiden Cox put it, in NSF's usual "'let's have it clearly understood that we are not trying to direct you' fashion."[110]

The broad-based, systemic nature of the instructional interventions BSCS developed provide the most obvious parallel to the curricular approach of the MIT group. Echoing Zacharias, Grobman wrote, "It was visualized from the first that such a study necessarily must involve not only the content of courses but the entire teaching-learning process of the students at all ages."[111] Perhaps the key component of the new biology curriculum was the laboratory. Seeking to address the criticisms from the scientific community regarding the decline in quality of laboratory teaching, BSCS developed two distinct approaches to such instruction. The first consisted of newly developed, but more traditional lab activities integrated into the daily classroom lessons. The more innovative tack sought to exploit the advantages that came with extended study of a single topic. This could only be accomplished, however, using what BSCS termed "lab blocks," periods of sustained laboratory activity designed to occupy the attention of a class of students for five to six weeks. Ten of these blocks were planned, though it was expected that a single group of students "would usually take not more than one block a year." Both types of laboratory instruction involved the development of new and inexpensive lab apparatus.[112]

Complementing the lab materials was a comprehensive film program. Initially established as a separate project of AIBS, the films were quickly integrated into the high school curriculum program.[113] The films were to serve three distinct functions for the biology course. The first was simply a move toward efficiency, similar to that undertaken by PSSC. "It was felt that the function of the laboratory work strictly as observation might be fulfilled by better organization of the course and the use of modern teaching methods including motion picture and television techniques."[114] "Technique" films were also developed to demonstrate important lab procedures. This was viewed as essential by the biologists to instruct not only the students, but the teachers as well, who themselves had little "first-hand" experience with such work. The third type of film was promotional, describing the history and philosophy of the BSCS program.[115]

The textual components of the course included a series of informational pamphlets for teachers on a variety of important biological concepts and a monograph series for students, which consisted of paperback books "similar to those being produced under the auspices of the PSSC."[116] The centerpiece of the program, however, was to be the textbook itself. Its development was handled for the most part by the Committee on the Content of the Curriculum (CCC), whose members included two later additions to the BSCS team: Marston Bates, a biologist from the University of Michigan, and John A. Moore, who chaired the committee, from Columbia University's Department of Zoology. A companion handbook for teachers was to be prepared by the Committee on Teacher Preparation headed by Joseph Schwab from the University of Chicago.

Initially the Content Committee planned to prepare material for a single integrated textbook. However, the BSCS Executive Committee decided instead to push for a series of stand-alone, single-topic monographs, which would provide teachers with needed source materials without dictating any particular sequence of presentation—a decision that did not sit well with the members of the Content Committee and was particularly exasperating to both Bates and Moore. Moore raised his objections at the third Steering Committee meeting in January 1960.[117] At issue was the perceived competency of high school teachers to use conceptually discrete monographs to present a coherent year-long course in biology. The problem of teacher quality was a significant factor that contributed to the highly prescribed, systems approach to curriculum development in both the BSCS and PSSC projects. After some intense Steering Committee debates, Moore reported that "the 'separate monograph' notion was abandoned."[118] The Committee had come to the consensus that high school teachers were "simply not sophisticated enough to utilize such materials." Teachers needed, as Grobman later stated, "a package" that they can use to substitute for current textbooks, which are relied upon as "a very strong crutch."[119] In creating these curriculum "packages," one BSCS writer later explained, they had "tried to make it as difficult as possible for teachers to teach other than the way we want it taught."[120] Wishing to avoid the "authoritarianism and totalitarianism" that a single textbook would imply, however, the Steering Committee agreed instead to produce the independent units and then have various subcommittees arrange them into three distinct textbook versions, each with a particular thematic emphasis.[121] NSF, concerned as always with the appearance of dictating school curriculum, was pleased with the report of this multifaceted approach.[122]

In later years, BSCS held itself up as an exemplar of the proper approach to curriculum reform. The "curriculum study," Grobman wrote, was to be hailed as the "new arrival on the American educational scene."[123] Of course the roots of the pattern and organization of this work in biology, as we have seen, were firmly grounded in the earlier efforts of the physical science elite that rose to power after the war. The National Science Foundation did their best to recruit the top scientists and, through the well-established contract model, ensured that BSCS would operate in an ostensibly autonomous fashion under the auspices of AIBS. Less explicitly, Foundation officials used the influence of their oversight and the example set by the MIT physicists to shape what they saw as the most effective approach to curriculum reform in science—comprehensive, systematic reform, which relied on scientific expertise in content, instructional technology, and operational design to support both teachers and students in developing a greater understanding of science.

## *Sputnik* and the One Best System

A good deal has been written about the impact the launching of the Soviet satellites had on education reform in the United States. By now it has been well established that efforts to rework high school science curricula were already underway before *Sputnik*.[124] Indeed, as we have seen, the origins of the postwar curriculum reform movement go back to the late 1940s, born of the ferment generated by critics such as Mortimer Smith and Arthur Bestor. The social factors at play—both ideological and technological—trace a continuum, with varying degrees of emphasis, throughout the Cold War period. McCarthyism highlighted the ideological component of the communist threat to American ideals in the early 1950s. The role of *Sputnik* was to shift the balance of public attention toward the technological/military threat the Soviet Union posed. The effect of this shift on federal education policy was to legitimate the scientists' perspective regarding the best approach to educational reform.[125] Whatever its philosophical appeal, science provided the foundation on which the technological and military strength of the United States rested and, given the crisis atmosphere across the nation, scientific know-how in research and education was called upon by the federal government to meet the challenge.

The Russian success of placing this rather modestly-sized aluminum sphere (just under two feet in diameter) in orbit around the earth made a disproportionately large impression on the American public. Despite well-known plans for launching a U.S. satellite as part of the upcoming International Geophysical Year of 1958 and assurances from the Eisenhower administration that the Soviet launch came as no surprise, public anxiety over Soviet intentions and capabilities shot up precipitously.[126] For some the military implications were unclear. Zacharias had to explain to one acquaintance that an orbiting platform from which the Russians might rain nuclear warheads down upon the United States would be an unlikely next step in the Soviet space program due to the tremendous inefficiencies such a plan would entail.[127] Nevertheless, the satellite did provide hard evidence that the Soviets possessed the rocket technology necessary to launch an intercontinental ballistic missile. Geographic isolation would soon be a thing of the past and, as one writer noted, "the American voter . . . is going to realize in his bones, that we can be hit and we can be hurt."[128] For most Americans, what *Sputnik* hit hardest was U.S. scientific prestige. The clear technological supremacy the country enjoyed at the end of World War II had seemingly evaporated overnight.

A flurry of articles in the popular press laid the blame for this American defeat at the door of the nation's schools. Following on the heels of the earlier curriculum controversy and the ever-worsening building and teacher shortages, such criticism could easily be expected. The latest "crisis in education," according to *Life* magazine's five-part "urgent '*Life*' series," was the product of the soft American curriculum. The functional life-adjustment curriculum that had been the target of the academic traditionalists in the early 1950s, despite its enfeebled state, was subjected to a renewed vigorous assault.[129] The *New York Times* later reported that even President Eisenhower urged that parents and educators abandon the "educational path, that rather blindly they have been following as a result of John Dewey's teachings."[130] Critics repeatedly contrasted the unfocused American course of study with the rigorous, science-based Soviet curriculum. Navy admiral Hyman Rickover in his book *Education and Freedom* was particularly scathing in his critiques of American schooling. The implications were clear: American students needed to buckle down if they hoped to match their Soviet counterparts.[131]

In a series of speeches to the American people, the president sought to reassure a jittery public that, despite the Soviet feat, the nation's security was well in hand. The thrust of Eisenhower's public response to the orbiting satellite centered on highlighting the importance of a long-term commitment to maintaining American scientific preeminence well into the future as the bedrock of the country's military strength. As a first step toward this goal, he appointed MIT president James Killian to the newly created office of Special Assistant to the President for Science and Technology.[132] Killian's job was to advise Eisenhower on scientific matters with the help of the Science Advisory Committee, which moved from the Office of Defense Mobilization to the White House and now reported directly to the president via Killian. Perhaps the most important symbolic move, however, in securing the proper place of science in society—and placating the nearly hysterical critics of public schooling—was the revitalization of science education. While acknowledging the importance of education in all academic subject areas, national security and political reality required that science be "single[d] out for special consideration."[133]

The actions taken by Eisenhower followed closely the recommendations of his science advisors, which came in a closed-door session on October 15.[134] Killian and the new President's Science Advisory Committee (PSAC), which included Waterman as an ex-officio member along with key PSSC insiders Land, Purcell, Rabi, and Zacharias, had Eisenhower's complete attention.[135] As a result of that meeting, the president stated publicly that the scientists wished to enlist his support in "awakening the United States to the importance and indeed the absolute necessity of increasing our scientific output of our colleges and universities" and expressed his faith in whatever plans for science education the scientists might develop, willingly accepting his role as the point man in promoting their cause. "I will do it," he declared, "because I believe exactly what they said."[136] Waterman took this opportunity to send Eisenhower a personal copy of the first draft of the PSSC textbook as an example of the kind of work that was essential to "stimulating our progress in science education."

"If this movement is as successful as we believe," he wrote, "it may well set a pattern for other sciences to follow."[137]

Eisenhower planned to ask Congress for $1 billion to improve the nation's scientific schooling. Congress, with both houses controlled by the Democrats, pushed for considerably more.[138] There was little doubt one way or the other that a surfeit of federal dollars would soon be available for science; the only matter to be settled was deciding which governmental agency would be responsible for administering the resulting programs. There were really only two options for dispersal of the new funds. The Department of Health Education and Welfare (HEW), home of the U.S. Office of Education (USOE), had an obvious claim to direct any new educational programs. The National Science Foundation, however, possessed the necessary scientific expertise as well as programs already in operation. Waterman had always recognized the delicate nature of Foundation support for education, especially pre-college education. Along with Harry Kelly, he worked hard to maintain amicable relations with the Office of Education to avoid any potential conflicts.[139]

Following *Sputnik,* the competition for funds began to heat up. At a meeting of the Divisional Committee for Scientific Personnel and Education on November 14, Kelly, referring to the Office of Education, "voiced some concern" over what appeared to be "a narrowing interest in science and engineering in that organization."[140] Killian argued in favor of NSF taking the lead in any federal initiatives, and John Stambaugh, the president's education consultant, prepared a lengthy memo also making the case for NSF. In it he argued that the Foundation was likely to be more effective in its efforts because "its program has the support and the esteem of scientists." Stambaugh viewed the cooperation of scientists in this matter as essential. "HEW, and its Office of Education, does not have the prestige that NSF enjoys," and therefore its programs "simply would not get the support from scientists that NSF now has." If the bottom line was "to develop more scientists," clearly this was "the job of scientists."[141]

Compromise with respect to funding was inevitable. In September 1958 Eisenhower signed the National Defense Education Act, which provided $1 billion over four years to HEW to provide need-based loans and fellowships for college students, funds for the purchase of school laboratory equipment and foreign language development, and support for research on the use of educational media in the classroom, among other items.[142] Not to be left out, the National Science Foundation received an infusion of funds that nearly tripled its budget for fiscal 1959. These were used to expand the existing fellowship and institute programs and, most notably, the Course Content Improvement Program. The popular focus on school curricula reignited by *Sputnik* propelled the PSSC project to the fore as emblematic of the direction such reform should take. Zacharias's curricular work was featured in the *Life* series on education and made headlines in both the *New York Times* and *Time* magazine.[143] The hope that Waterman had expressed to Eisenhower regarding the pattern-setting nature of the physics curriculum came to fruition in the establishment of additional curriculum projects in chemistry, mathematics, earth science, and, of course as we have seen, biology.[144]

The disparity in funding between HEW and NSF was considerable, and somewhat surprising given the consensus within the administration regarding the lead role scientists should play in guiding reform. The billion-dollar appropriation HEW received with the passage of NDEA for the four-year period from 1959 to 1962 dwarfed congressional funds for NSF education programs over the same period nearly seven to one. Funding levels, in this case, however, were not the most reliable indicators of the direction of national education policy. Under the shadow of the Soviet satellite, intellectual authority clearly rested with the president's science advisors. The appointment of former ODM-director Arthur Flemming as the new HEW secretary in 1958, further assured the scientists of a measure of influence over the direction of programs in the Office of Education. In fact, not long after Flemming's elevation to secretary, Caltech physicist and PSAC member Lee DuBridge communicated to Flemming a "concern that some people have around the country" that the newly established NDEA programs will be "directed by the 'professional educationists,' as contrasted to the scholars, scientists, and other teachers in subject-matter fields." He suggested, somewhat predictably, the "appointment of an advisory committee of scientists" as one way "to avoid this danger."[145]

It is difficult to gauge the precise manner in which the scientists wielded their influence across the various administrative divisions. What is evident though, particularly in regard to curriculum reform, is the ascendancy of the scientists' modus operandi. The National Science Foundation, in tandem with Zacharias and the Physical Science Study Committee, not only set the standard for curriculum projects in the other sciences, but legitimized an operational approach subsequently embraced by officials at the Office of Education for nonscience subjects as well. In an address before a national curriculum conference held at the University of Minnesota in 1961, J. Boyer Jarvis from the Commissioner of Education's office praised the work being funded through NSF's Course Content Improvement Program. He indicated that the USOE intended to follow this model and "do for other basic subjects what the National Science Foundation has done . . . for science and mathematics."[146] Commissioner of Education

Sterling McMurrin even solicited advice from PSAC's newly formed Educational Research and Development Panel chaired by Zacharias. In one meeting, McMurrin commended the "enormous value" of the Foundation's work and asked the scientific panel specifically for guidance regarding "things to be done; how the Office should go about them; and in what priority" for curriculum reform in the "non-science fields."[147]

Technological mastery over the natural world was a coveted asset scientists brought to the table during the Cold War conflict in the 1950s. Real or perceived, the concrete threat of *Sputnik* tracking overhead lifted their status even higher. Zacharias and the scientists of PSAC and PSSC, already confident in the power of their disciplinary methods within their specialized fields, grew increasingly confident in the ability of those techniques and methods to solve nearly any problem, social or physical. As Nobel prize-winning physicist Edward Purcell later recalled, "we were at that time pretty much imbued with the idea that we could do anything if we started from scratch, and did it in a rational way, and applied all our technology to it and so on. It was, of course, a time of almost euphoria in that respect, because the wartime successes of high technology and physics applied to new problems were still very fresh in our minds."[148] The intellectual contest with the Soviets had provided the opportunity for scientists to try their hand at revitalizing the science curriculum; *Sputnik* prompted the Eisenhower administration and a large segment of the public to pin their hopes on the scientists' success.

## CHAPTER 5

# PSSC: ENGINEERING RATIONALITY

The technocratic approach to curriculum design articulated by the Apparatus on Teaching group at Woods Hole would have received unmitigated praise from the national security officials within the Eisenhower administration had the new curriculum materials been tuned solely to produce greater numbers of scientists. The imperatives of national security required a no more nuanced view of science education than this; increasing the absolute number of technically-proficient citizens was all that was needed. The chairman of the NSF's Division of Scientific Personnel and Education had little difficulty recognizing this. "The Government's interest," he observed, "is primarily in having the tools necessary for defense," and it was increasingly apparent to everyone, as another divisional committee member stated, that "science is [now] the most important arm of defense."[1] The needs of the scientific community, as we have seen, however, diverged sharply from those of the national security apparatus. Thus, with respect to NSF's curriculum reform program, what the government sought to purchase—primarily technical expertise—differed significantly from what Zacharias and the rest of the scientists at NSF were prepared to sell: science education designed to meet "the Foundation's highest ideal"— the "furtherance of research as a vital part of the intellectual, moral, and cultural strength of America."[2] Given the leverage scientists possessed (the belief that only they had the appropriate training necessary to produce more of their own), it was clearly a seller's market.

NSF officials realized that integrating science, and the scientific worldview, into the fabric of mainstream American culture required first and foremost addressing the persistent public misunderstanding of modern science. Here they ran smack into the rising tide of mass culture that characterized the 1950s. NSF director Alan Waterman laid out the rather timeworn problem in his talk at the Conference on the American High School held at the University of Chicago in the fall of 1957 just after *Sputnik*. In discussing the topic of "Science in American Life," he offered a telling anecdote: "A television comedian recently quipped: 'The Russians got the satellite up first because American scientists are busy testing tooth paste and administering hypodermic injections to fountain pens.'" He explained that "this attempt to be funny is a perfect example of the prevailing misconception that labels as science a wide variety of activities ranging all the way from large-scale technology to tricky 'gadgeteering.'" Waterman went on to complain that "the advertising industry not only spuriously identifies the scientist with a host of commercial products but often succeeds, in the process, in making him something of a figure of fun." From NSF's perspective, the role of the schools in addressing this less than desirable situation was straightforward. "If we are to realize the full potentialities of science in American life," Waterman declared, "we must recognize the need to educate the public to a more perfect understanding of the true nature of science and how it is fostered."[3]

This longstanding prescription for securing the proper place of science in society was, as expected, followed to the letter by Zacharias and his Physical Science Study Committee. Although Zacharias wholeheartedly embraced the systems engineering techniques of the Cold War defense projects, the curricular system he engineered was designed to deliver more than mere technicians. Technocratic means did not imply technocratic ends. His target was the literate public, those destined to become the leaders of industry and the nation; his goals were much broader. For science to assume its place as the foundation of American culture, the "central purpose of school education," as most scientists believed, "should be the development of the 'rational man.'"[4]

The Physical Science Study Committee worked diligently under the direction of Zacharias to construct a view of physical science that would advance this goal in the interest of science and the scientific community in the United States. To be sure, the physicists involved were committed to an accurate picture of the content and practice of modern physics, but it was a picture created with more encompassing social goals in mind. In addition to creating a basic understanding of modern physics distinct from its technological application, they sought to teach about the fundamental practice of basic scientific research and the conditions required for its most rapid advancement. The centerpiece of this curricular approach was an emphasis on the process of reasoning from empirical evidence, which was intended to counter the rampant anti-intellectualism and the growing public irrationalism in American Cold War culture. Nearly all the deliberations over the content and presentation of the new physics in the text, films, and laboratory activities focused on the goal of disseminating this particular view of science.

## Recasting the Subject Matter

The PSSC Steering Committee began its work by embracing the broad principles outlined in the new physics syllabus developed for the state of New York in the mid-1950s. The New York outline stressed the need to focus on the fundamental concepts of physics over the usual emphasis on the mass of factual information so often included in high school physics courses. The MIT physicists stood solidly behind this approach. Treating fewer topics in greater depth would be key, they believed, to helping students develop a deep understanding of the discipline. To make room for this more integrated, holistic picture of physical science, much of the traditional subject matter would need to be eliminated; the first items to be purged were the everyday examples of the technological applications of physics. The group felt that this move would not only create the necessary space for a more coherent course, but would also help eliminate the confusion over the difference between basic and applied science common among the general public.[5]

The physicists had surveyed the most frequently used textbooks in high school physics, chemistry, and physical science courses and were uniformly disappointed with what they found. All the books included "much too much material, use[d] words inaccurately, and fail[ed] to give emphasis to fundamentals as distinguished from examples."[6] The consensus at MIT was reinforced by the results of a study of high school physics textbooks chaired by Walter C. Michels in the spring of 1956, which was jointly sponsored by the American Institute of Physics, the American Association of Physics Teachers, and the National Science Teachers Association. Michels' committee unanimously concluded that of the 14 most commonly used textbooks, not one could be considered satisfactory.[7]

The root flaw in all these books, from the scientists' perspective, could be traced to the distorting influence of pedagogical theory on the intellectual content of physics. In attempting to appeal to students and teachers from a wide variety of backgrounds, the publishers had adopted a reductionistic approach to the presentation of content, breaking the subject down into discrete, piecemeal units. This, the publishers claimed, helped facilitate student learning in a difficult subject. The physicists noted, however, that it also allowed them to stock their texts full of a variety of pet topics in order to appeal to the widest audience of teachers. Each of these units was further embellished with examples of the everyday applications of the science covered to increase student appeal. Rather than making the subject interesting and easy to learn, however, Michels insisted that in so fractionating the discipline the publishers had merely "succeeded in making physics into a very complicated subject," effectively sacrificing the subject to pedagogy.[8]

The heavy reliance on everyday examples, though consistent with the student-centered educational philosophy of the time, was particularly objectionable to the PSSC physicists. One staple of existing textbooks, for example, was the description of how the physical principles of phase changes are exploited for the process of home refrigeration. A student could find nearly every other modern convenience, from the telephone to the automatic transmission, disarticulated and displayed as well.[9] This emphasis on technology, the physicists felt, obscured the "general principles" of the discipline. And everyone agreed that the "fine typography and . . . technical advances in printing" the publishers used were wasted on content that did nothing to present the modern view. "One triple transparency overlay," they were amazed to find, served only "to clarify the details of a power shovel."[10] The factual knowledge presented as physics in these commercial textbooks was essentially "indistinguishable from the technological advances that have been based on [that] . . . knowledge."[11] It was no wonder the public remained unclear about the distinction between science and technology when high school courses themselves contributed to the confusion. Thus from the beginning, Zacharias and his colleagues were steadfast in their decision to exclude technological applications and examples from "household" physics from the new curriculum. This they accomplished, as they stated in their first annual report, "with little or no regret."[12]

With the technology question decided at the outset, the PSSC group used its three-day December meeting with representatives from around the country to decide what exactly would go into the new course. All agreed that the unity of physical science should be evident throughout the curriculum. A concept as simple as the conservation of energy, for example, might be introduced and then followed through its many applications in the fields of electricity, optics, or atomic physics. There was less agreement at the meeting, however, regarding what topics and approach would specifically be used. Some argued for taking a very small part of physics and treating it with "the detail and rigor one finds in a good course in Euclid"—helping students painstakingly reconstruct the conceptual structure of physics from first principles. A second group pushed for presenting the subject in its historical context, an approach in which science would be cast as "a counter force to superstition and blind faith."[13] Zacharias noted that during this "stormy" December meeting the "walls bulged and the ceiling went up and down." But, he added, "we ended up with a reasonable outline of the way we wanted it to go."[14] The group had decided to walk the middle path between the two views, committing to a course that focused broadly on "the role of atoms in the physical world."[15]

The argument for the unity of science was a strong one. The PSSC curriculum had been initially conceived as a two-year course in physical science that would have deliberately blurred the boundaries between the traditional disciplines of physics and chemistry.[16] Indeed, at the December meeting many of those present thought it "inadvisable to erect an arbitrary division between them."[17] The resulting physics course reflected this commitment, crossing over into chemistry whenever necessary to provide a complete picture of atomic theory for students. The more ambitious plan for a two-year course, however, never got off the ground. When asked by NSF why chemists had not been more directly involved with PSSC, Zacharias replied that he had tried to enlist their participation in the project but found that, as he diplomatically phrased it, their "interest and enthusiasm" did not match that of the physicists.[18] It was a situation he clearly was not happy with. Looking back, Zacharias attributed the chemists' unwillingness to participate to professional jealousy. In light of the dominant role physics played during World War II and continued to play during the Cold War, it is, perhaps, not surprising that the chemists demurred, reluctant to see chemistry serve as handmaiden to physics in a high school physical science course. When given the opportunity by NSF a few years later to develop a course of their own, the chemists apparently were able to muster the appropriate enthusiasm.[19]

Making a distinction between science and technology, as we have seen in the argument for the establishment of the National Science Foundation, was crucial to securing the proper public support for basic scientific research. While Zacharias and his group were constructing the new physics curriculum with this distinction clearly in mind, various science advocacy groups were preaching the same message to a broader public. Taking advantage of the renewed interest in science following the Soviet venture into space, NSF released a report entitled "Basic Research: A National Resource." In this report, transmitted to President Eisenhower just two weeks after *Sputnik* and right before the Conference on the American High School, Waterman opened by describing the profound difficulty people have understanding the concept of basic research. "The picture most of us have is distorted by the emphasis given in the news to the exceptional, unusual, 'newsworthy' aspects of new inventions and developments often lacking reference to the fundamental work without which these could not have been made."[20] Promoting greater public understanding of the nature of such fundamental research was necessary, the report unabashedly stated, first "*to establish conditions more favorable to the continued growth of basic research,*" and, second, "*to achieve a greater flow of funds for basic research.*"[21] This NSF document was followed in the spring of 1959 by a symposium on basic research sponsored by the AAAS, the NAS, and the Sloan Foundation. Symposium presenters included familiar speakers on this topic (Alan Waterman and James Killian) as well as those less formally associated with the issue, such as President Eisenhower. All recognized the crucial need for a better informed public.[22]

The Physical Science Study Committee was doing its part. The textbook it produced the following year, true to the group's intent, described few if any physics-based technologies students might encounter in the course of their daily activities. In the opening pages of the book, the authors made a point of distancing their work from its applications. The "physics so important in everyday life," they explained, "is left to a professional group of engineers, far removed from the general study of physics."[23] The only technological examples the physicists did include were those related to the instruments and machines of scientific research itself, such as the cyclotron, the bubble chamber, and the mass spectrograph. The elimination of everyday examples was a radical departure from the science textbook of the preceding decades, in which automobiles and refrigerators stood alongside power shovels in stimulating student interest in the science of day-to-day things.[24]

## The Process of Science

The conceptual relationship between basic science and technology was rather interesting from a political standpoint. While the scientific elite found it necessary to trace clear distinctions between the two ideas for the purpose of advancing the cause of basic research, those responsible for securing public funding (NSF administrators, for example) were careful to make sure that the connection between science and technology, however deliberately attenuated, was never completely lost. When dealing with the practically-minded Congress, there was no better justification for continued support of basic science, as Vannevar Bush had demonstrated, than its capacity to produce impressive technological applications.[25] Though the argument was an easy one to make, back in the quiet courtyards of MIT it was not one that the physicists themselves felt was all that compelling. In their view, the ultimate value of scientific research was in its exemplification of the transcendent human effort to understand the natural world.

Speaking about the PSSC project at an MIT alumni-day symposium on the future promise of science, Zacharias set the new curriculum squarely within this more humanistic point of view: "To many people, science is regarded as a kind of seed corn to be grown and hoarded, in case at some future time, we need to grow a new crop of technology—to give us *more*—more *things* . . . more control over the forces of nature, more mechanisms for the besting of our rivals and more devices for the confusion of our enemies. But to many of us who work in science, in laboratories, who work at theory, or teach science to younger men, we think of science as a part of our culture."[26]

Zacharias's reference to "our" culture could be taken as an acknowledgement of the isolated intellectual world in which the scientists worked, a world perilously at odds with the mass culture of American society. More likely he meant simply that all facets of modern society—the broader culture—had become permeated with the fruits of scientific thought, and what was needed was for the general public to be brought to a greater level of awareness of this fact. Both to some extent reflected the conditions of the time. The primary goal of PSSC was to remedy this state of affairs by formally admitting science to the humanistic foundations of Western civilization.

The curriculum reform effort at MIT was, in fact, one manifestation of a much larger trend by scientists to eradicate the distinction between science and the humanities in the postwar period.[27] (Harvard's general science curriculum initiated by Conant after the war was another.) The recent course of societal advance, they insisted, demanded a reconsideration of what knowledge was fundamental to the proper operation and management of that society. The humanities had long been central to the liberal arts deemed essential to the education of the literate public; but the world had changed during the twentieth century, and changed dramatically since the end of World War II. "Science [has] become inextricably interwoven with our daily life," wrote Elbert Little, the executive director of PSSC. It has "left an ineradicable mark on our modes of thought and our methods of choosing between alternate courses of action."[28] The traditional curriculum was no longer up to the task of preparing those who would govern, do business, or educate in the new postwar world. As MIT provost Julius Stratton insisted, "a liberal education must . . . be relevant to time and circumstance. . . . It should fit [men] to perceive and comprehend the great issues of *our* time, the forces that are shaping *our* destiny." The liberal arts, he argued boldly, should thus be reconstructed around the "one great, unifying force working in our age, and that is science. We must turn to science for the *lingua franca* of modern men and find in science the vehicle of modern thought."[29]

Little made this case specifically for the new high school physics course. Science, he argued, quoting the humanist intellectual Charles Frankel, "is an example *par excellence* of a liberal art." It had the power, quite literally, to liberate mankind, not only in the sense of its ability to free man from the drudgery of the material world or to defend against totalitarian enemies through its technological applications, but, more importantly, in its ability to "carry men beyond the foreground of their experience" and enlarge "the dimensions of human choice by acquainting men with the alternate possibilities of things." This was the central lesson about science the physicists wanted students to learn. To the members of PSSC, the process of scientific inquiry was in and of itself "a 'final good,'" something that could "give stability and direction to the rest of our lives."[30] The importance of this latter power of science was another of the key reasons they eliminated everyday technologies from the course—to highlight the intellectual value of science independent of its material value to society.[31]

As a practical matter, communicating the liberating power of scientific thought to high school students required a good deal more than just banishing technological examples from the textbook. The PSSC group was aware that it needed specific curricular elements that would illustrate the nature of scientific research and help students experience firsthand, to some degree, the process of scientific inquiry. Here was where most traditional textbooks resorted to the perfunctory first-chapter description of the so-called "scientific method" for students first to learn and then practice through repetition. This standard treatment of scientific thinking was viewed with contempt by working scientists across the board. To the PSSC physicists these introductory chapters were full of nothing but "false ideas" about the methodology of science.[32] The inclusion in biology textbooks of the common lock-step prescription for arriving at scientific knowledge struck BSCS writer Marston Bates as so absurd he wondered aloud to a colleague whether "the fellows who teach biology in our country really believe the crap about 'scientific method' with which they uniformly start their textbooks."[33] Richard Paulson, the NAS executive secretary of the Committee on Educational Policies and later an NSF staffer, thought that such opening chapters were best omitted altogether "so little do they show any real effort to find a conceptual framework . . . or appreciation of what is involved in scientific investigation beyond 'the method' set forth in Deweyite simplicity."[34]

The association of the scientific method with John Dewey, and therefore with modern educational theory, immediately identified it as something detrimental to the achievement of real intellectual goals in the eyes of scientists. The "method" as advanced in the educational field can be traced back to Dewey's analysis of the psychological process of problem solving described in his book, *How We Think* (1910).[35] It consisted of five steps: "(i) a felt difficulty; (ii) its location and definition; (iii) suggestion of a possible solution; (iv) development by reasoning of the bearings of the suggestion; [and] (v) further observation and experiment leading to its acceptance or rejection."[36] Dewey laid out these steps (though he insisted their order was not fixed) as a guide to intelligent action. The process was scientific to the extent that the worth of the problem solution arrived at was measured by its ability to overcome the initial difficulty in practice, that is, as it was experimentally tested. Indeed, he argued that the methods of science, broadly conceived, should serve as the foundation for all decision making, especially that exercised in the social realm. This schematic "scientific method" gained wide currency among professional educators in teachers colleges and professional education associations throughout the country. Its rigid, algorithmic application to problems of any sort, much to Dewey's dismay, became its defining feature in the hands of the education establishment.[37]

From the educator's perspective, this generalized, stepwise procedure was ideal precisely because it was not limited to scientific problems. This made it a perfect fit for the life-adjustment emphasis then in vogue. Scientists might not have been so concerned were it not for the educators' drive to make this method the primary learning outcome of science instruction. Some educators insisted that the "development of competence" in its use should transcend in importance all the "other objectives of science education."[38] The emphasis on this seemingly inflexible, simplistic method, scientists believed, served only to distort the true process of science.[39] They felt it minimized the complexity of scientific research and glossed over the unique conditions necessary for such research to flourish. Real scientific work required men of powerful intellectual talent, who, far from proceeding in any stepwise fashion, creatively deployed a wide range of sophisticated problem-solving strategies in their pursuit of knowledge about the physical world.[40] Many scientists viewed research as a craft, which required a level of expertise that took years to acquire. The method, if one were to insist on such a term, emerged from the intersection of a large body of mostly tacit practical knowledge with the more formalized theoretical structures of science in the consideration of a discipline-specific problem. Moreover, much of the success and power of this approach came from the ultimate simplicity of the phenomena, not the simplicity of the process. One could hardly expect, as educators apparently did, that these discipline-specific techniques would be equally applicable to the complex social phenomena that underlay the problems in students' daily lives.[41]

Zacharias and his colleagues sought to infuse their new physics curriculum with multiple opportunities for students to gain some understanding of what they felt real scientists did, to get a close-up look at the kinds of problems *scientists* solved and the way in which current conceptual knowledge was brought to bear in their solution. Their understanding of this process was largely implicit. No one on the Steering Committee operated from any well-articulated or systematic account of scientific inquiry. Most were content to embrace physicist-turned-philosopher Percy Bridgman's rather loose definition of scientific method, which was "nothing more than doing one's damnedest with one's mind, no holds barred."[42] Zacharias and Philip Morrison, at least, were certain that the essence of this approach consisted of the logical relationship between knowledge claims and the empirical observations from which they flowed. Zacharias was highly regarded as an experimental physicist, and his vision of the discipline was grounded in this professional experience. For him the work did not center on abstract theoretical considerations, "physics . . . was what you do either in a laboratory or by observation."[43]

Zacharias realized that a high school physics course was not going to teach students the intricacies of advanced scientific research; the curriculum, despite the manpower concerns of the ODM, was not intended to prepare students to embark on scientific careers.[44] He believed that a well-developed course, though, could provide students with an appreciation of the complexity of scientific practice and, more important, could instill an understanding of the rational foundation of all scientific work. The question Zacharias hoped to get students to ask themselves at all times was "how do you know?" What was your "basis for belief" in any assertion about how the world works?[45] Recognizing the evidentiary basis of knowledge in physics—of reliable knowledge generally—was the most important lesson high school students were expected to leave with. Far greater than technical content knowledge or specific laboratory skills, here was one aspect of science that the physicists believed truly had broader social applications. Ironically, this argument for the value of a scientific approach (in the sense of examining the empirical warrants for any belief) to social problems was one with which Dewey would have heartily concurred.

Many scientists felt that such skills were sorely lacking in America during this time. "We can all agree—the world being what it is," Paulson complained, "that dispassionate looking for facts, intelligent skepticism, willingness to accept the tentativeness of conclusions, readiness to see the hazards of premature acceptance of universals, and so on, are far from common."[46] As we have seen earlier, scientists were highly sensitive to their professional standing in society in the 1950s, a standing they felt was threatened by a growing irrationality among the masses that continued throughout the decade. The outbreaks of public hysteria over issues such as fluoridation and fallout, congressional disrespect for basic research, and most notably the public zeal for security restrictions and the concomitant effort to expose communist sympathizers all flew in the face of what, scientists felt, any reasonable person would deem rational behavior. It was these manifestations of the antiscience culture of America that most concerned the scientific elite.[47] In the spring of 1956 the internationally respected physicist I. I. Rabi made headlines claiming that the "American esteem for science and scientists was now 'lower than it has ever been in this century.'" Disheartened, he noted the irony that even "as the importance of science in the country increases, its dignity seems to be diminishing."[48]

Rabi's sentiments were certainly shared by Zacharias, his close friend and advisor, as well as many of the physicists working on reforming the science curriculum at MIT. Zacharias had felt firsthand the sting of overzealous efforts to uncover disloyalty among intellectuals. The highly politicized nature of the anticommunist investigations of the early 1950s made high-profile colleges and universities with close ties to defense research, such as MIT, ripe targets for attack. Zacharias had served as the chairman of the Committee on the Responsibilities of Faculty Members, established in 1953 by the Institute after charges of subversion had been leveled at one of its professors. He had also testified at the Oppenheimer security hearing in 1954, during which he had himself been accused of belonging to a subversive cabal of physicists that had sought to undermine national security.[49] Such harassment led him to consider the possibility of leaving academic employment altogether in 1955. In preparation for such a move, he and Jerome Wiesner, who later became President Kennedy's science advisor, jointly founded a technical consulting firm as a fall-back position in case they were forced to leave the Institute. As Wiesner explained, the establishment of the firm "was almost a direct response to our interaction with Joe McCarthy."[50] In one instance, two of McCarthy's more infamous associates "came around, looking to get into the personnel files of the Lincoln Lab and RLE [Research Lab in Electronics], and they got turned down flat." Some in the MIT administration, upon learning of this lack of cooperation with government representatives, were agitated enough to lead Wiesner and Zacharias to believe that their jobs might be in jeopardy as a result of their transgression. The matter was eventually dropped without incident.[51]

The experience of PSSC author Philip Morrison told a similar tale. Morrison had worked on the Manhattan Project during the war. He had, in fact, performed the final assembly of the bomb that was dropped on Hiroshima and soon after toured both cities devastated by the new weapons. In vivid testimony before Congress he described the horrible human toll of the atomic blast at Hiroshima. Although he had not been a member of the Communist Party since 1942, Morrison, a student of Oppenheimer's, was continually harassed by red hunters throughout the 1950s for his unfailing support of the peace movement and nuclear disarmament. In 1949, *Life* magazine singled him out as a subversive threat in an article on eminent American communists. This was followed by a variety of well-publicized reports on the menace he posed to society because of his political views. An entire section was devoted to him in a Senate subcommittee report on the "subversive influence in the education process," which ran in *U.S. News and World Report* in 1953. This negative press rankled and embarrassed board members and alumni of Cornell University, his employer at the time, and eventually led its president to threaten him with termination unless he dissociated himself from left-wing, though perfectly legal, political causes. Although not fired, for a time Morrison was repeatedly passed over for promotion to full professor despite his brilliant record as a physicist.[52]

Political attacks such as these reached endless heights of absurdity. At one point a federal judge categorically asserted that "the younger generation of scientists, particularly in the field of physics, has succumbed to communist propaganda." He made a point of exculpating those in practical fields, such as "chemistry" and "engineering," targeting instead "younger persons 'engaged in *pure science*.'" Individuals working in the most abstract domains of knowledge, he claimed, were especially susceptible to being swayed by communist ideology.[53]

These personal encounters with the hard edge of the Cold War political realities of the time that Zacharias, Morrison and, no doubt, other physicists experienced played an important motivational role in the drive to reform the high school physics curriculum. They powerfully reinforced the general professional concerns scientists had over the fate of science in America. In describing what prompted his involvement in the project, Morrison expressed his belief that public education had been very much against the spirit he thought was good for science. Furthermore, he explained: "I was oppressed by the feeling of the early fifties that science and intellectual reason itself were not being given a fair chance in the schools and in public life." It seemed to him simply an "important public duty to do something about that." Given the situation at the time, the high school courses, particularly physics, "were strategically a very advantageous place to go." The need for a more rational citizenry was equally apparent to Zacharias. His decision to lead the curriculum reform project was based, as he stated, on "*deep* political reasons," which stemmed from the "Joe McCarthy era." These served for him as the ultimate justification for the course. Based on his experience during the war, and after seeing "the American public . . . being *molded* by Joe McCarthy," it was perfectly clear to him, looking back, that "to get people to be decent in this world, they have to have some kind of intellectual training that involves the topics: observation, evidence, and basis for belief." These were the reliable intellectual tools that could combat the irrationalism of the time; the primary purpose of PSSC curricular system was to deliver those tools effectively.[54]

## Physics on Film

The film component of the PSSC course was crucial to the advancement of the physicists' goals. Not only did it provide a means to alleviate some of the problems related to the shortage of qualified science teachers, as Henry Chauncey had argued some years before, but more importantly the films gave the physicists direct access to the classroom, to both students and teachers. Here was an opportunity to demonstrate the nature of scientific reasoning, to open a window into the daily work of physics that would allow students to see, "real scientists dealing rigorously and respectfully with real problems, and deriving not only satisfaction but excitement from the intellectual pursuit in which they are engaged." In this way the films would function to bridge "the gap that so frequently arises between a science . . . as it is taught in school, and the same science . . . as it is actually carried on by its practitioners," thus normalizing the public image of scientists in the process.[55]

Of all the components of the curriculum, however, the films proved to be the most difficult to produce. The early conceptual work began in the fall of 1956. Decisions about the approximate length of each film, the total number, type of media to use, and the nature of the integration of film and text were made in short order. General guidelines regarding the visual content of the films were settled without controversy as well. To convey the appropriate intellectual weight of scientific work, the films would avoid using "pretty girls, unessential details in animation, and unnecessary action." "Only the necessary props" would be included "to avoid distractions."[56] But once shooting started in the spring of 1957, things slowed down considerably. Zacharias, who was in charge of the film program, quickly learned that "the production of a movie—the conversion of an idea onto celluloid—was a very difficult process."[57] By summer he had succeeded in making only a short pilot film that demonstrated the pressure of light, and he was the first to admit that it had "obvious defects."[58] In the fall of that year, the Steering Committee agreed to bring in Encyclopedia Britannica Films to speed the process along with the understanding though that the physicists would have the final say over the results.[59]

Creative differences erupted almost immediately. The central point of disagreement was over the nature of the film narration. Britannica's standard procedure for making educational films was to shoot the phenomena and experimental apparatus in whatever sequence was desired and then to add a voiceover narration describing the visual sequence and explaining the salient scientific points. From a production standpoint, the elimination of people on screen made the film much easier to script and execute. It could often be done in pieces rather than all at once. Zacharias believed, however, that this approach, though more expedient perhaps, compromised the overall goals of the project. "The disembodied voice," to him, "seemed very unattractive," and contributed to an "attitude of awe and wonder in place of simple realism."[60] The only acceptable way Zacharias could see to capture the essence of scientific work was to put the physicist himself in the film. This Britannica was reluctant to do. After nearly a year of disagreement and unmet expectations, PSSC and Britannica dissolved their association.[61]

The question of whether physicists would appear on screen was no small matter for Zacharias and the rest of the Steering Committee. One of the central commitments of the new curriculum was to demonstrate how scientists went about their work. Visual recreations of physicists in action, they felt, were the most effective means of accomplishing this goal. Not just any physicist would do, however. The development and deployment of atomic weapons at the close of World War II had saddled scientists, in some ways, with a less than desirable public image. While they had always seemed to exist just outside the social mainstream, their contribution of these weapons of mass destruction to the world arsenal added a new dimension to the stereotype—that of a little madness.

The physicists at MIT were well aware of this image: the scientist as "a long-hair off in an ivory tower, lost in clouds of thought no ordinary mortal could hope or want to understand," according to Zacharias; or, as a AAAS poll of high school students found, a muttering, potentially subversive, brain with no social life, who bores his wife and children.[62] These were hardly the images on which to build a renewed faith in scientists and the scientific worldview. Seeing real scientists on film, the Steering Committee felt, would go a long way toward correcting these misconceptions. In the PSSC films, scientists would appear "not as a disembodied intellect but as a normal, active, and occasionally fallible human being."[63] "The inevitable small blunders of hand and speech" that were certain to occur without professional actors, Zacharias insisted, would be "humanizing."[64] Perhaps most important to Zacharias in his self-described role as "casting director" was to thoroughly Americanize the image of the scientist, "to show that a physicist was *not* a Hungarian with a briefcase talking broken English but the Ed Purcells of this world and the J. R. Zacharias's, somebody who spoke English with no accent, who was one of the boys."[65] Even the PSSC teacher's guide made a point of reminding teachers to let students know that physicists were very much "like other Americans," that most "marry, have children and belong to PTA's; some play golf and bridge and watch westerns on TV."[66]

Though the films were an important conduit for rehabilitating the image of the scientist, their primary function was to model explicitly the process of scientific reasoning, reasoning that at its most basic level might carry over into the real world. Such reasoning, which focused on the tight relationship between theory and evidence ran, of course, throughout all the elements of the curriculum. However, the demonstration of this process by physicists in the films was "expected to set the tone and level for the entire course."[67] Zacharias laid out his dialectical vision of how science worked: "We do experiments from which we make a theory. The theory suggests new experiments. We perform the experiments. We modify the theory usually as the result of the experiments, and so on. I call it right foot, left foot, right foot, left foot."[68] The most important goal for every film was to get the "logic lines" between the empirical data and the theory "crisp, clean, clear, and charming." In this way it would be possible, as Zacharias described in his initial memo outlining the project to Killian, to build the entire conceptual structure of physics from a continuous series of experiments.[69] Given the quality of science teachers in the classroom, the demonstration of this type of reasoning was as important for them as it was for the students. Zacharias went so far as to claim later on that the films were made only "ostensibly for the students." They were developed mostly for "teacher training," so that "the teachers could see how the arguments ought to go, how the logic lines ought to be."[70]

Zacharias's pilot film, *The Pressure of Light,* provides an excellent example of the PSSC attempt to capture the process of scientific thinking.[71] An early version, which featured an MIT engineering professor as the on-screen scientist, was previewed by NSF officials in November 1957. Though it received "a great deal of favorable comment," Zacharias was dissatisfied with the level of enthusiasm the engineer was able to project.[72] In the end, Zacharias felt that only he himself was capable of demonstrating for students the proper amount of excitement for the apparatus featured in the experiment, which he had devised and constructed.

The film opens with a close-up of a handgun, which is then used to shoot a tin can sitting a short distance away. The film's tone, its emphasis on careful reasoning in place of transmitting physics content, is set by the opening remarks. The camera cuts to Zacharias, standing in a rumpled suit, who explains in a casual manner: "Now this film isn't going to be about guns and bullets. It's just that in the last few lessons, we've pictured light as if it were a stream of particles of some sort." He goes on to describe how this particle model of light accounts for a variety of phenomena. The scene shifts to each as he enumerates them: "Remember that reflection, and re-fraction, the fact that light moves in straight lines, all seem to fall into place. It's true that we had to make a few extra assumptions, but the picture isn't too bad." The analogy to bullets is then developed further: "Let's take something the theory suggests, something we know about bullets and see if it's true about light. What happens when you shoot at a tin can? The bul-let doesn't just stop and fall to the ground. For a short time it exerts an im-pulse. Now, if light behaves like a stream of bullets, it ought to do something of the same sort."[73] Thus, Zacharias sets up a situation where the theory, in this case light as a stream of particles, allows one to make a prediction—that light must exert a pressure. All that remains is for Zacharias to confirm in some way that this is indeed the case.

In a series of carefully sequenced experiments, Zacharias tries to pro-duce the push that the theory predicts should occur. After attempting un-successfully to topple a carefully balanced mirror with light from a powerful automobile spotlight, Zacharias turns his attention to a piece of apparatus long believed to demonstrate this effect: Crookes' Radiometer (familiar to most as an evacuated glass bulb containing a weather-vane-like pinwheel made up of four small diamond-shaped panels, each having one reflective side and one blackened side). When light shines on the radiome-ter, the pinwheel "just whizzes around," purportedly the result of light par-ticles striking the flat surfaces of the vanes. What is important for instructional purposes, though, is how well the phenomenon matches the specific prediction of the theory with this particular apparatus.

As Zacharias demonstrates, careful examination of the situation reveals some surprises. He begins by walking the viewer through the process of "how Crookes reasoned this out." Past experience shows that a particle that strikes an object and bounces off exerts a greater force than the same particle that strikes an object and sticks to it. Therefore, with respect to light particles and the vanes of the radiometer, Zacharias explains, one would expect that light striking the shiny surface of the vanes, being re-flected, would exert more force than the light striking the blackened side of the vanes, where it is absorbed. Based on this reasoning, the pinwheel, when placed in a beam of light, should spin with the shiny sides trailing. The camera shifts to a close-up of Zacharias shining the light on the ra-diometer again. But as he points out, "it's going around with the black side trailing. So it's wrong." Clearly the evidence, plainly visible to a careful observer, fails in this case to support the theory in question, it simply "doesn't add up." "Now nature isn't wrong. The theory's wrong," Zacharias explains. "Instead of demonstrating the pressure of light, he [Crookes] must have been demonstrating something else. . . . He was convinced that light must exert a pressure. But clearly he hadn't proved it."

The rest of the film continues in this deliberate stepwise fashion, deducing from theory what one would expect to observe, eliminating confounding variables (in the case of the radiometer, the thermal effects from the residual amount of air present in the glass bulb was held to be responsible for the spinning of the pinwheel), and then comparing theory predictions to experimental results. With a more efficient vacuum and a much more delicate apparatus, Zacharias in the end succeeds in demonstrating the effects of light pressure on a fine piece of foil suspended by a thin quartz fiber.

The film concludes with a recap of the reasoning process he had gone through. Zacharias states: "We had a theory that light was a stream of particles, sort of like bullets. We know that when bullets strike an object they exert a momentary force. And so we reasoned that if the bullet theory of light is a good one, then a beam of light should push on any object that it strikes. And then, after some difficulties, we found that light does exert a force. And so the bullet theory does indeed predict the correct behavior of light." The emphasis throughout the film is on how one goes about warranting scientific knowledge claims, not according to some formulaic process, but in the ostensibly real way physicists themselves supported and argued their claims in practice. There were well over 50 films produced by the mid-1960s, each following the lead of Zacharias's *Pressure of Light,* demonstrating the way physicists went about their work.[74]

### Into the Laboratory

While the film program provided a window onto the world of the physicist, films by their very nature were limited by the passive role in which they placed the student. Zacharias understood that "only so much physics can come through the eyes and ears; the rest must pass through the hands."[75] In the PSSC system, the shortcomings of any one curricular component were balanced by the strength of another. In this case, the classroom laboratory provided the hands-on experience with natural phenomena that the MIT group felt was essential for a complete understanding of the rational basis of scientific thinking. Ultimately the labs served two functions in the PSSC program: they provided the opportunity for students to appreciate the empirical grounding of scientific knowledge firsthand, and at the same time, in their simplification of the experimental epistemology of physics, they sought to demystify the complexities of postwar laboratory research in order to help the public understand scientific inquiry as a cultural endeavor—man's creative search for meaning.

Even prior to his involvement with PSSC, Zacharias had been convinced that the laboratory was where all science instruction should begin. Year after year MIT undergraduates had arrived in his classroom eager to assimilate current theoretical developments in physics, rarely appreciating or even seeing any need to grapple with the empirical warrants for that knowledge.[76] Zacharias saw this uncritical acceptance of knowledge on the basis of scientific authority as undesirable from a broader social perspective certainly. But, much closer to home, he felt that this theory-first mentality undermined the foundational basis of physics as science. The authors of the PSSC Laboratory Guide noted this tendency to esteem theory over experiment, which was often reinforced by high school labs that were set up merely to confirm previously taught physical laws or principles.[77] "Although science is often pictured as wholly the work of mathematicians and theorists, such is by no means the case." They explained that "without carefully contrived laboratory techniques . . . the science of physics (and chemistry) would never have developed beyond the dream stage."[78] For Zacharias, who operated squarely within the positivistic spirit of the time, it all came down to what happened in the lab, and there the "physicist leaves no . . . stone unturned in his search for physical truths."[79] The PSSC curriculum was built on this rock of empirical observation— the laboratory experiments were designed to "pave the way for the reading of the text" and students were to be assured by the teacher that, whatever else they might find, "nature is not wrong."[80]

With the experience they gained in the laboratory, students themselves would be able to explore the evidentiary basis underlying scientific beliefs; no longer would they be "merely a matter of what the students had read or what they had heard."[81] The proper coordination of the labs with the text, the physicists felt, would enable students to develop a deeper understanding of the dialectical march from experiment to theory and back again—the right foot, left foot Zacharias had spoken of. For him, it was always the right foot of empirical evidence that determined the left foot of theory in physics. There was no denying that theory provided the conceptual tools necessary to probe nature, but nature, in his mind, ultimately provided the corrective to theory.

Creating classroom activities that conveyed the physicists' preferred conception of laboratory research was not exactly a straightforward task. Many scientists of the time recognized the difficulty described by the chemist and former Harvard president James Conant in the introduction to the *Harvard Case Histories in Experimental Science* (the material developed for his undergraduate Natural Science 4 course). For the layperson, he wrote, "modern science has become so complicated that today's methods of research cannot be studied by looking over the shoulder of the scientist at work."[82] World War II, as we have seen, had radically transformed the American research laboratory, and the continued patronage of science by the defense establishment during the Cold War had pushed postwar research even further toward the "big science" archetype—large-scale research projects built around complex instruments and apparatus of truly industrial scale directed by teams of researchers that numbered into the hundreds. Among some scientists there was a growing sense of nostalgia for the simpler days of laboratory bench work. Historian Peter Galison writes that for others these changes were even more disconcerting: "Many experimenters saw the new modality of research as threatening the very foundation of their separate existence as physicists."[83]

The PSSC group, it seems, was driven more by the pedagogical requirements of the task at hand than by any existential crisis. In either case, the members of the group made a conscious effort to restore the intimacy of the relationship between the "scientist" and the physical world—at least in the classroom lab. Though they understood the reasons for the war-induced transformations of the scientific laboratory, many of the changes, they felt, obscured the fundamental aspects of experimental inquiry. The essence of this work for them was not to be found in the sophisticated instruments or large-scale apparatus scientists used. "Contrary to popular belief," they informed the student, "a laboratory need not be a sprawling glass-walled building bulging with giant electronic computers, huge cyclotrons, instrument panels, and row upon row of apparatus-laden benches." Real scientific work—what really mattered—required only the intellect and perhaps some "equipment assembled from readily available odds and ends."[84] The point that Zacharias and his colleagues sought to make was that the practice of science was located within the individual and not in any particular place or external laboratory apparatus the scientist might manipulate; it was a way of *thinking* first and foremost. "Many of the really important advances in physics," they explained, "have been made not in formal laboratories, but wherever the physicist happened to be when he was confronted with a problem."[85]

The intellectual aspects of laboratory research were thus foregrounded as much as possible in the lab activities developed. One way they accomplished this was to de-emphasize the role of manufactured apparatus and laboratory technology. For the most part, the physicists effectively eliminated traditional lab equipment altogether. In its place they sought to have the students construct their own apparatus from scratch, using various combinations of sheet metal, wooden blocks, soda straws, match boxes, and other everyday items. From these materials they were to produce micrometers, optical range finders, and microbalances sensitive enough to weigh a fly's wing. One of the stated reasons for this cardboard and twine approach was to reduce the cost to local districts that might be interested in adopting the curriculum. Given the financial strains school systems were under during the 1950s, this limiting factor was certainly wise to consider. But, more important, was the pedagogical reason that "complicated apparatus is apt to obscure the basic simplicity of the subject under investigation." It was hoped that when a student performed experiments in a PSSC classroom, the function of the mediating laboratory apparatus would be transparent, thus allowing clear logical connections to be made from phenomena to theory, from evidence to knowledge claim. This, Zacharias felt, would instill in the student an appreciation for the importance of the "basis for belief" he so frequently spoke of, and also in this way, "the role of the scientist become[s] more meaningful to him"—the intellectual role, that is.[86]

Furthermore, the construction of laboratory apparatus from common household materials, the physicists expected, would erase the distinction between the home and the lab. As explained in the laboratory guide, this approach "helps to prevent a split between the student's world and that of science. . . . In fact these two worlds are the same."[87] To encourage this outlook on science, the guide included numerous suggestions at the end of each lab for additional experiments that might be carried out at home. "There will be little the student will use in his school laboratory that he cannot duplicate in his garage or cellar," Zacharias noted.[88] Suggestions such as these were intended both to portray the excitement of an endless intellectual frontier in science as well as to integrate, at some level, methods of scientific reasoning into the everyday world of the students—to clear a path for science to make its way into the cultural mainstream.

The first draft of the student laboratory guide was produced in 1958 and consisted of three volumes. The first contained a good deal of background information on various shop techniques such as the proper manner in which to cut and bend glass tubing, how to work with sheet metal, and the use of different resins and cements. The rest of the volume provided schematics and directions for the construction of the various pieces of apparatus that would be used throughout the course. The second volume of the lab guide began with a series of activities in which students were to explore the use of the instruments and analytical tools they had constructed, becoming familiar in the process with the range and the level of accuracy in measurement they might be able to expect with each. This part of the course was closely tied to Zacharias's identity as an experimental physicist, to his experiences fashioning field-ready radar equipment at the Rad Lab during the war. He expressed great enthusiasm for having students fabricate, and then test and calibrate their own instruments. "The building of apparatus," he insisted, "is as much a part of physics as its judicious use, and the procedure is justifiable on these grounds alone."[89] This explicit emphasis on the measurement of phenomena and theoretical questions concerning the nature of measurement itself, which pervaded the curriculum, no doubt reflected the operationalist orientation that guided the practice of most experimental physicists in the mid-twentieth century—an orientation in which theoretical concepts were meaningful only to the extent that they could be defined by specific operations (measurements or calculations) performed by the scientist.[90] The remainder of the labs paralleled the textbook, treating key conceptual ideas central to the understanding of modern physics.

Early returns from the classrooms piloting the PSSC curricular materials suggested that the physicists had perhaps been a bit too enthusiastic in their commitment to the student creation of their own lab materials. Teachers complained that the work often took an inordinate amount of time, and they sensed that students felt that, once the instruments were built, there were few real activities for which they were needed.[91] Classroom reports also indicated that "many girls have experienced some difficulty in the construction and complicated manipulation of the apparatus they have been asked to build."[92] Eventually Zacharias and the Steering Committee settled on providing classrooms with sets of preassembled laboratory kits to eliminate the apparent burden of construction both teachers and students experienced. This, they felt, was a suitable compromise that would buy more time in the laboratory while retaining the simplicity of the apparatus that was central to the conceptual understanding of experimental practice.[93]

## Education for the Age of Scientists

After four years of intense work, the first edition of the PSSC physics course made its official classroom debut in the fall of 1960. All things considered, it was a well-crafted instrument to be wielded by the nation's high school teachers on behalf of the American scientific community. It is important to recognize, however, that although the new curriculum was without question informed by professional interests, these interests need to be understood in their broader social and political context—specifically within the context of the "Age of Science" ideology advanced by the scientific elite. The relationship between the social role science was to play in postwar America and the goals of science education were articulated most completely in the statement on education released in 1959 by the President's Science Advisory Committee's (PSAC) Education Panel chaired by physicist Lee DuBridge. It is in examining this document that we can begin to fully understand the educational aims of Zacharias's new high school physics course.

Zacharias and his colleagues at MIT played a central role in drafting the PSAC "white paper" on education, which was eventually titled *Education for the Age of Science.*[94] The other members of the Education Panel were comfortable deferring to Zacharias on this topic, mostly as a result of what they perceived as his recently acquired expertise in the field of education.[95] Zacharias certainly had given a lot of thought over the previous few years to what the appropriate goals of science education should be. Along with many scientists of the time, he had become convinced that good science instruction above all else should work to eliminate two dangerous and ever-widening gaps in society—those between intellectuals and nonintellectuals and between humanist scholars and scientists. The divergence of interests among these various groups was obvious to nearly everyone. Despite its increasing importance in society, science clearly was not "as widely appreciated and sought out as other cultural elements—like art, music, history, literature." PSSC members noted that "many who are not going to be poets, painters, or musicians spend time to train their ears and eyes, to learn terms, to understand techniques and methods." Few people, they felt, made similar efforts to understand or appreciate science.[96]

Although the Education Panel in the end cast the problem as one of *mutual* understanding, the individuals sitting on PSSC and PSAC viewed the difficulty, not surprisingly, in a more unilateral sense, as the public misunderstanding science.[97] It is also clear from their various statements and discussions that the "understanding" they sought quickly shaded into appreciation, respect, and even deference. Questions of cultural status and value were front and center in the minds of the scientists and, ironically, they looked somewhat longingly at the social position scientists enjoyed in the Soviet Union, where, if not for the stifling pressure of communist ideology, the scientific management of society might have proven itself effective. Edwin Land explained to Eisenhower, during his meeting with PSAC following the launch of *Sputnik,* that the Russians "regard science both as an essential tool and as a way of life." As a result they "are teaching their young people to enjoy science."[98] In contrast, the people of the United States possessed from the beginning a "traditional disregard for abstract thinking," and, given the exploding consumer culture of the late 1950s, there was the sense among scientists that increasing material comforts were slowly eroding any hope that affairs of the mind might make their way toward the cultural center of American society.[99] In a speech before the Women's National Press Club in 1958, James Killian emphasized this point: "One cannot fail to ask whether we Americans in our drive to make and acquire things have not been giving too little attention to developing men and ideas."[100] In a country with such a strong spiritual heritage, it seemed somewhat paradoxical in the eyes of many scientists that the life of the mind would be so denigrated, whereas in the Soviet Union, Killian noted, "a system of government based on materialism has found a way to bestow its highest awards on men who deal in abstract ideas."[101]

The Education Panel looked seriously at measures that would generate greater public status for scientists, such as the recognition of role models and the awarding of prizes for intellectual excellence.[102] One administrative consultant saw the task as one of creating the appropriate "prestige symbols so that the next generation will not disparage 'things of the mind.'" Only in America, he noted, do you have a society that "puts Jane Mansfield's breasts, or Dietrich's legs above the value of an Oppenheimer or a Rabi."[103] Whatever immediate measures might be conceived to get the country on the right track, all the Panel members agreed that the long-term answer lay in education. At the high school level this meant the revision of course content, the development of new learning aids and supplementary reading materials, and an emphasis on do-it-yourself science labs—all the components of curriculum system developed by the PSSC group. These were the ways, PSAC believed, by which "our education can be strengthened so that it will more fully meet the requirements of this age of science, and best serve the nation at a time when the security of the Free World and the defense of human freedom are inescapable responsibilities of the United States."[104]

But all this to what end? The goals of the PSSC program, as we have already seen in great detail, were to create a more "accurate" image of the nature of postwar scientific research and, more generally, to instill the "the power of scientific reasoning and logic" into every educated person.[105] A public that understood the rational basis of scientific thought, the physicists reasoned, would be able to resist the destructive materialism of American culture and the anti-intellectual demagoguery of politicians like McCarthy; both were viewed as social forces highly detrimental to the advancement of scientific knowledge. PSSC member Walter Michels summarized the situation when he stated: "The future rests not so much on the number of engineers and scientists produced each year as it does on the intellectual climate in which they work, and this, in turn, will be determined by our success or failure in making physical science a significant part of general education."[106]

The effort on the part of scientists to develop a social and intellectual environment favorable to scientific advance—an environment defined by NSF as one in which the scientific community received ample public funding yet retained complete political autonomy over the direction of research—should not be interpreted as narrowly self-serving, though it may appear so at first glance. The world had been transformed by the products of science during the war, and Zacharias along with the rest of the physicists at Los Alamos and the Rad Lab had been the ones who had brought about that transformation. There was, in their estimation, "no way to turn back the clock." The growing stockpiles of American and Soviet nuclear weapons testified to this fact like no other. The United States had come to depend on the expertise of scientists as the Cold War raged, a situation to which there appeared to be no alternative. The Education Panel concluded that "the people of the United States, on the most practical grounds, *must accept and support these propositions.*" A technocratic, expert society had emerged from World War II and for such a society to survive, scientists believed, their profession needed to be understood, respected, and appreciated. It would not be enough in this age of science for the public to merely "applaud and reward [scientists] or their contributions while still thinking of them as useful strangers, dimly understood and more feared than admired." What was needed was a "widespread dedication to and respect for learning in all fields, and a deep understanding between the public and the experts."[107]

The fact that the typical senior-level course in physics might reach at most only one-quarter of the total high school population was of little concern to Zacharias. That 25 percent of students represented the future elite of the country—the "lawyers, businessmen, statesmen, and other professionals."[108] In an expert society, these were the nonscientists in positions of power who needed to appreciate and understand science. More important, scientists felt it was they who must recognize the necessary role of science in directing the new technocracy. Despite the spectacular contributions science had made to society beginning with World War II, it had failed, in the words of the AAAS, "to attain its appropriate place in the management of public affairs."[109] The PSSC curriculum project had been developed, in part, to provide what was lacking in American society: a broad-based rational approach to public policy. The stakes were sufficiently high in the Cold War struggle with the Soviet Union that the scientists felt they had to begin at the top. "We could not aim at the people who most needed help in education," Zacharias recalled years later. "We had to establish a first-class collection of stuff for the intellectual elite of the country, no question."[110]

## CHAPTER 6

# BSCS:
# SCIENCE AND SOCIAL PROGRESS

L ate in the summer of 1960 at the University of Colorado, BSCS Chairman Bentley Glass addressed the cohort of teachers that had volunteered to teach the first drafts of the biology course that had been hastily put together over the previous few months.[1] He began his talk by recounting the revolutionary progress that biology had made since the beginning of the century, noting the various developments in the fields of genetics, biochemistry, and medicine. With almost messianic fervor he sought to capture the awe of the assembled teachers and enlist their support in paving the way for the revolutions in biology to come. "What will biology be in the year 2000?" he asked rhetorically. "We will probably learn," he began modestly, "how to increase the human life span." He grew bolder: "I would suspect that by 1990 biologists will have learned how to create some simple forms of living organisms . . . and that geneticists will have learned how to replace defective genes with sound ones." Glass also predicted that biologists would learn how to conduct artificial photosynthesis and, above all, "man will certainly have learned to accelerate his own evolution in a desired direction." The question was "what direction will he desire"?

The teachers were the key to this wondrous new world that awaited. But the road would not be easy. Glass explained to those assembled that there would be opposition: "Some of this opposition, as you know, will come from the S.P.C.A. or the Humane Society in your neighborhood, from persons who think that it is not necessary to examine the secrets of life." There would also be opposition certainly "from anti-evolutionists, perhaps a dying breed, and from others—who are more common and more to be feared than anti-evolutionists—who are simply 'anti-science' in point of view." To combat these forces of ignorance, the teacher must help the "citizen of this country, the man in the street," to "learn what science really is." This, he insisted, would be "our primary task."[2]

The volunteers would not be heading into the classroom empty-handed. The new BSCS curriculum, in its nascent form, would provide the structure and guidance they needed. Glass's talk provides nearly a complete snapshot of the underlying social and political agenda the biologists sought to pursue through their NSF-funded curriculum project. In his reference to the expected battle with the antiscientific element among the public, Glass aligned BSCS with the developers of PSSC, who worked to combat the rising irrationalism that threatened the continued autonomy and support of science. Glass's explicit references to technological applications and the social promise they held, however, revealed a split from the educational agenda of the physical scientists. As much as they both shared in the common scientific ideology of the time, the biologists' perception of the needs of science—of their professional needs in particular—did not completely square with that of the physicists.

There were three important themes in the materials produced by the Biological Sciences Curriculum Study: evolution, human progress, and scientific inquiry. These themes, among others, reveal the specific concerns that prompted the biologists to become actively involved in influencing the public perception of biology, concerns that stemmed primarily from the dominant position the physical sciences held after the war. Envious of the widespread prestige and influence the physicists enjoyed during this period, the biologists sought to secure for themselves a similar position in the new technocratic society that had emerged. Thus they used their new course materials to showcase the importance of the biological sciences to social progress, casting biology as a discipline equal to the physical sciences in its power to control the forces of the natural world. At the same time, they set their discipline apart by emphasizing the irreducible nature of biology's evolutionary perspective and the unique contribution it could make to ensuring the ultimate survival of humanity.

Unlike the Physical Science Study Committee, which was directed by a relatively small steering committee that generated a single first-year physics course, BSCS was both organizationally and intellectually more diffuse. The BSCS project employed a significantly greater number of writers and consultants throughout the development of its first set of materials. The project also committed itself to developing three different textbook approaches to the teaching of biology, which were labeled by color: the yellow version emphasized cellular biology, the green version focused on ecology, and the blue, molecular biology. The themes and key ideas described below, however, ran throughout the project as a whole. While some of these are more apparent in some versions than others, the ideas, in every case, were of central concern to the biologists on the Steering Committee and to American biologists generally in the late 1950s and 1960s.

## Biology in the Age of Science

The similarity of the goals and rhetoric of BSCS to those of PSSC was, in some respects, quite remarkable. There was no love lost between the biologists and physicists involved with the respective projects. Estimations of incompetence were voiced by each group for the other on more than one occasion.[3] Despite their petty differences, these two groups shared a vision of the important role science education needed to play in modern society. The postwar "age of science" ideology had a hold on the consciousness of the American scientific community that transcended disciplinary boundaries. But along with the endorsement biologists gave the general technocratic aspirations of American scientists, they devoted their energy to addressing the specific place of biology in society as well. The needs of their profession emerged at the beginning as an important consideration in the development of the new high school curriculum.

Of all the members of BSCS, Steering Committee Chairman Bentley Glass was, perhaps, the most vocal in stumping for biology education. Given the amount of writing he did and the pervasiveness of the "age of science" rhetoric within the scientific community, it is not surprising that nearly every argument made for PSSC, and science education more broadly, could be found in his assorted public statements. The integration of science and the humanities, for example, was a strong theme in his work. He went so far as to suggest, as other scientists had before him, that all the standard school subjects be reoriented to serve the advancement of science. Learning to express the logical structure of thought in high school English, he insisted, should take precedence over "purely descriptive, narrative, or 'creative' writing." He urged that history as a school subject be expanded beyond its treatment of past human activity to include the history of the earth and the subsequent evolution of animal life as part of its proper scope. Only by "fully and more consciously relat[ing]" these subjects "to the place of science in human life," could science truly become the core of a modern liberal education.[4]

Recasting the liberal arts around science in this way was an essential step, Glass believed, to safeguarding science from its longstanding enemies, "the authority of tradition, of religion, or of the state." Here he highlighted science's "most intimate relation to human freedom," echoing the associations of science and democracy so forcefully stated by the scientific community during the height of the Red Scare in the early 1950s.[5] Furthermore, he stated even more explicitly than the President's Science Advisory Committee the importance of educating the public to appreciate the nature of the scientific society in which it lived. This was a matter on which the survival of the nation depended. He affirmed "that a nation of a microscopically few scientists molding and altering people's lives, and a populace uncomprehending, superstitious, and resistant to the novel ideas of the scientist while blandly accepting the technological fruits of those very ideas, . . . cannot endure." Unless the public became "truly scientific in spirit and in endeavor," we would ultimately, he proclaimed, "face oligarchy, and eventual collapse of our form of civilization, our way of life."[6]

Public understanding of science was important, to be sure. But while the physicists might have been satisfied with the goals outlined above, for Glass the additional matter of disciplinary balance was a point that needed to be addressed when considering the meaning of scientific understanding. He was conscious of the "lowly esteem in which the biologist [was] held in comparison with . . . the chemist or the physician," though "physicist" was likely uppermost in his mind. He felt this situation was largely the fault of the biologists themselves, who had never bothered to build "strong professional organizations . . . to speak for their interests and their principles."[7] The new BSCS curriculum was one way they might begin to redress that imbalance. The study's director Arnold Grobman made this same point when he insisted that "the basic science taught in the schools should include biology in proper perspective with and in relationship to the other fundamental scientific disciplines. Every educated person should be able to count among his philosophical resources an understanding of evolution, of genetics, of energy relationships, as well as the principles of optics and mechanics and what makes a *Sputnik* go."[8]

The apparent marginalization of biology can be traced at least back to World War II, when the military first began to rely heavily upon the techniques of the physicists. The failure of the military to likewise "make use of the professional abilities of biologists" was a glaring insult to those in that field.[9] As research in the physical sciences increasingly paid military dividends throughout the Cold War, disparities in funding and status grew. During the 1950s, for example, federal government agencies consistently favored the physical sciences and engineering over the life sciences.[10] Combined funds for the support of research projects and research facilities in biology lagged considerably behind such funding in the physical sciences. For its part, the National Science Foundation gave biological research a prominent place in its budgetary plans, though inequities remained in the fellowship program, where awards in the "hard" sciences outnumbered awards in the life sciences nearly three to one.[11] NSF director Alan Waterman acknowledged the reality of the overall funding imbalance in a talk before the Boston Society of Biologists in 1954. Although the Foundation was committed to funding basic research in all the disciplines equally, he explained, the fact remained that "financial support of considerable size . . . is obtainable only when the public realizes that the need is great." This was the case "for defense research and medical research," and, in this respect, "the physical sciences and the medical sciences have the same advantage over the social sciences and the pure biological sciences as, say, weapon research has over pure mathematics." The heaviest support, he conceded, was most likely to occur "when the field can be dramatized, as in the case of Radar, atomic energy and most feared diseases."[12]

Waterman's point about the public perception of the various sciences was clearly grasped by the biologists on the NAS Committee on Educational Policies. In their discussions in the spring of 1955 on how best to improve the country's biology education, questions of competitive advantage with respect to physics were central. The overall goal of the Committee was to improve the position of biology as a professional discipline in the United States. This required, some argued, both developing greater "professional solidarity" among biologists as well as increasing the number of students going into biology. When it was suggested that the Committee back programs to increase the absolute numbers of all scientists as a means, thus, to increase proportionally the number of biologists, objections were raised that such an approach might actually be detrimental. "It may be that other disciplines like physics would attract [the] most able among the additional group, and the overall [effect] might be a general lowering of the competence of people going into biology." This comment pointed to the widespread feeling that "biology [had] not yet been so presented rightly enough to be a challenge to enough young people with the best minds." The Committee decided in the end that if there were to be a priority rating, first priority should go to projects that will improve education generally—to get "more staff, more money, more buildings. . . . Second priority should aim at improving the challenging character, the competitive position of biology." And finally, to avoid giving any advantage to the physicists, "the third priority would go to efforts aimed at promoting the status of science at large in the community." The biologists knew they had work to do. The Committee chair had made clear at an earlier meeting that "physics, chemistry, [and] mathematics already [were] putting out much material, with no reference to biology. We don't want biology to lose out."[13]

The appearance of earth-orbiting, artificial satellites could not have come at a worse time for American biologists. The public outcry over the loss of U.S. technological superiority these devices seemed to signal led them to believe that they might very well lose out to the physical scientists entirely.[14] Concern over the direction of the Eisenhower administration's science policy in light of these events prompted the American Institute of Biological Sciences' (AIBS) executive director, Hiden Cox, to plead the biologists' case to James Killian, the president's science advisor. "Biologists are genuinely concerned," he began, "that science will fare badly in the present climate of confusion and in some quarters, hysteria." He insisted that to bring America back to scientific preeminence would require "the efforts of all scientists in all disciplines. . . . Science must advance on all fronts. Physics cannot advance as rapidly as it should if biology and chemistry lag. These interrelationships between scientific disciplines are understood by most scientists but I do not believe that the public yet realizes how firm these bonds are." One way to ensure the proper direction of policy in this area, he suggested, would be to expand the Science Advisory Committee "to allow the effective participation of biologists in deliberations."[15] Killian, of course, saw through all the talk about balance and the interconnectedness of scientific disciplines. The point of "all this hub-bub," he noted, was that "biologists want their place in the sun of the new space age—and to make sure of this [Cox] wants them well represented on the Science Advisory Committee."[16]

The primary problem biologists faced in attempting to reposition themselves on the scientific political spectrum was one they themselves recognized—they lacked a coherent disciplinary and professional identity. One University of Chicago zoologist observed that "in its fast advance, biology tends to break up into isolated columns, splintering into ever more subdivisions of specialization and particularization."[17] Marston Bates, prior to joining BSCS, noted that in all the efforts to "sell" biology, to do something about its place in the "peck-order" of science, "we talk about [it] as though it were a real thing, an important area of thought. But what do we find when we start out to look for biology? Mostly we find a queer hodge-podge of conflicting, overlapping, self-important special sciences."[18] Even a casual look at textbooks, where one might expect to find some more general picture of the discipline, was "very discouraging to anyone looking for integration, synthesis, in biology."[19] This was certainly the case with two of the most popular high school biology textbooks of the time— Moon, Mann, and Otto's *Modern Biology* and Smith's *Exploring Biology*— each of which was viewed as little better than a compendium of assorted biological facts.[20]

Members of the BSCS Steering Committee shared this assessment of their profession. Many felt that part of the reason for the lack of student following in biology as a field of study was the subject's inherent incoherence. BSCS director Arnold Grobman acknowledged that "its challenges are somewhat less clear than those of chemistry and physics. Its fundamental principles often are obscured by the emphasis that is placed upon comparisons between groups of organisms . . . or upon terminology and classification schemes." As a result, "some students do feel that biology is a diffuse, uncorrelated science which is uninteresting and, very likely, unimportant."[21] One member of the Steering Committee warned of dire consequences unless something was done. Without recasting biology as a unified discipline, he insisted, "it is my belief that we will run the danger of having the students as a group simply turn away from biology as a field of interest and then, as a concomitant of this, the physicists and chemists coming in from the back door will preoccupy the field and the biologists will be, as a profession, reduced to a small, ineffectual service group."[22] Glass agreed, "We can never have successful biology courses until we have true biologists who refuse to draw limits within the organic world."[23]

### Defining a Discipline

When the matter of determining the appropriate content of a high school biology course was taken up at the first BSCS Steering Committee meeting in February 1959, the chair, Bentley Glass, immediately directed the Committee's attention to "the problem of subdividing biology into areas." Disciplinary fragmentation was a foe not easily vanquished. Initial plans for a unified course in biology, as we have already seen, quickly spiraled into three distinct versions—yellow, green, and blue—each with its own subdisciplinary biases. But despite this concession to diversity of approach, the BSCS biologists insisted that there needed to be, underneath the variable presentations, a foundational view of biology as a whole. The desire was to locate this view in a set of key themes, which would, as Glass explained in a letter to the New York Times, "weave all aspects [of biology] together into a coherent, meaningful, and challenging unity." These themes would thus map out for high school students the place of biology in the vanguard of the scientific age and, at the same time, highlight its unique, irreducible features that distinguished it from physics and chemistry.[24]

The members of the Steering Committee had high expectations for effecting significant changes in the public's view of biology as a science. Much of this enthusiasm came from an appreciation of the large sweep of the school-age population that biology instruction traditionally touched.

Although there was some discussion about developing a twelfth-grade course, the benefits of leaving biology in tenth grade, where it was taken by "eighty percent of American students," seemed to outweigh the added depth that might be achieved with fewer students in a more advanced course. The choice was clear and BSCS, "anxious to have as many high school graduates as possible be scientifically literate," fixed its aim on tenth grade. One memo, playing up the positive side of the large tenth-grade audience, stated that the new curriculum would be the chance for biologists to measure up "to their opportunity to present to the student-citizen what he must know to be able to lead a satisfying and productive life."[25]

The decision to develop a "less sophisticated course" for a larger cross section of the American public, consisting primarily of 15-year-old students, required careful consideration of issues of accessibility and coherence.[26] "The main emphasis" of the curriculum was "to be on concepts and principles with reduction of technical terminology,"[27] reported the first issue of the BSCS newsletter. Conventional biology textbooks were littered with technical vocabulary—words often introduced only once, never to be seen again. At a meeting of the Committee on the Content of the Curriculum, a key BSCS subcommittee, John Moore reported that "someone presented the startling information that high school students are confronted with more new words in current biology courses than in their foreign language courses."[28] Bates felt that the multiplication of terms was part and parcel of the tendency toward disciplinary fragmentation. The goal of the new books would be to "keep technical terminology to a minimum aiming at half of [that found in] the current texts."[29] Glossaries were eliminated as well in favor of weaving the definitions of important terms into the narrative of the text.[30] The biologists hoped this might sidetrack the rote memorization approach to learning. Some "teachers are fiendish creatures and will make their students memorize incredible amounts of detailed information," observed one member of the Content Committee.[31]

The simplification of the vocabulary, of course, was directed toward the greater goal of communicating the unifying themes of the curriculum. Following Glass's request at the first meeting, it was agreed that the high school course be infused with nine key ideas: the nature of scientific inquiry, the intellectual history of biological concepts, genetic continuity, regulation and homeostasis, complementarity of structure and function, behavior, relation between organism and environment, diversity and unity, and evolution. Of the nine themes, the first two were singled out by the executive committee for fullest development.[32] These—the nature of inquiry and the historical development of biological ideas—were designed to place biology in the company of the other great scientific disciplines of the twentieth century. The remaining themes were, as Grobman noted, "related to the content of biology, the actual structure of biological knowledge," and served to demarcate biology from physics and chemistry.[33] Crosscutting these themes in the biologists' framework were seven levels of organization, starting with the molecular, passing through the individual, and ending with the world biome. These levels and themes, over the course of the numerous BSCS committee meetings, came to be referred to as the "warp and woof" of biology. "A course could be designed," Grobman stated, "by weaving these threads together to form the general fabric of biology."[34]

Taken together, the themes were intended to provide the intellectual framework in which to situate a unified picture of biology, which would help students "conceive of biology as a science, and of all the sciences as tested reliable methods of gaining objective knowledge."[35] Such an image contrasted sharply with the biology presented in existing textbooks. While physics and chemistry might have been distorted in the past as a result of pedagogical compromises or even plain ignorance, biology as a school subject had a long tradition of being deliberately crafted to meet perceived social needs, to teach hygiene or human physiology to the masses, for example.[36] Biology, more so than the hard sciences, was also susceptible to public censorship in areas where it intersected with cultural mores and beliefs. As part of an earlier study on biology teaching in America, Glass had found that the very topics BSCS wished most to emphasize—inquiry, evolution, biological principles—fell at the bottom of the average teacher's list of the most important topics, displaced by the perennial favorites, health, physiology, and human heredity. What formal biology was included consisted mainly of taxonomy and classification.[37] Following in the footsteps of PSSC, the biologists sought to "close the widening gap which more and more threaten[ed] to separate the classroom and the research laboratory in the biological sciences."[38] The goal was to present biology as a research discipline, rather than merely a collection of health tips for better living.

The members of the Steering Committee were steadfast in their commitment to an accurate depiction of modern biological research in the curriculum. For too long, they felt, biology had been distorted by external pressures. Contemptuous of school administrators who "don't feel the subjects in school have changed that much," Joseph Schwab insisted that the Committee "should not talk about what the schools want to do. As scientists our function should be to promulgate the ideas and concepts of our profession."[39] Three important areas that were crucial to communicating their view of biology quickly emerged as potential flash points—evolution, the use of live animals in the classroom laboratory, and human reproduction.

After some exploration of the issues involved, Grobman and the rest of the Committee resolved to stand firm. The general approach in all three areas was to follow Moore's suggestions for the treatment of human reproduction. He counseled that "the approach to the problem should be as biologists giving an adequate explanation of the material and then being careful not to modify our position on the basis of a small minority opinion."[40] With respect to evolution in particular, the Committee had decided firmly that "the topic would be fully represented." "There should be no compromising or pussyfooting in this area," Grobman stated emphatically.[41]

Glass strongly supported these efforts to maintain the intellectual integrity of the discipline. Perhaps more than any other member of the Steering Committee, he was convinced of the destructive power political and social pressure held for science. Glass was active throughout the 1950s defending academic freedom in science from such external threats. During the height of the Red Scare in 1953, he had his fingers on the pulse of the scientific community as editor of both AAAS journals, *Science* and *Scientific Monthly,* where key debates over scientific freedom and government oppression took place.[42] He was also heavily involved in the American Association of University Professors' response to the firing of faculty members who were associated, directly or indirectly, with the Communist Party. Glass, in fact, chaired the AAUP committee and cowrote the report on all of the university anticommunist cases of the 1950s. And as president of AIBS, he fought against the highly mobilized antivivisectionists, who sought to curtail or even eliminate experimental research on living organisms.[43]

For Glass, academic freedom and science existed in intimate association with one another. This association remained a personal and professional touchstone for him well into the late 1950s, if not for his entire career. In the fall of 1957, not long before BSCS got underway, Glass wrote: "Science has never, can never, prosper for long except where the mind is free. In the development of scientific freedom the struggle for the broader freedoms of speech and association now in our Bill of Rights, and the struggle for academic freedom in the universities, have been essential."[44] Such thoughts were never far from his mind. During one Steering Committee meeting in 1960, when the topic of external sponsorship of BSCS curriculum material was raised, Glass jotted down on his notepad some guidelines: "Two principles of intellectual freedom which must be maintained in sponsored publications: 1) The BSCS must have the right to choose the author . . . 2) The sponsoring agency may supply material to the author . . . but must in no case dictate or supervise his presentation."[45] These beliefs were, no doubt, voiced repeatedly to the entire Steering Committee during his tenure as chairman.

Though the biologists were quite ready to present an uncompromised vision of their science to the American public, success developing materials to reflect that vision remained elusive. Despite clear instructions going into the first writing conference held in the summer of 1960, BSCS writers struggled in their efforts to develop and integrate the nine guiding themes into the various textbooks.[46] The shortcomings were apparent to the Steering Committee. A number of members, for example, felt strongly that the themes of "inquiry and history and evolution [were] not adequately woven through."[47] Compounding the poor treatment of the themes was the inability of the first writing group to sufficiently streamline the technical vocabulary. One external reviewer, after reading the initial drafts, stated candidly that "the three texts vary in their awfulness but all show the signs of a classical descriptive isolated biology supposedly made modern by the injection of poorly digested abstracts of *Scientific American* articles." Of the BSCS biologists themselves, he noted caustically, "they are committed to their own folly and to the exercise of the students' memory with long strings of species of animals and now, to be modern, long strings of molecules."[48]

Subsequent revisions of the textbooks over the following summers, however, resulted in materials much closer to what the Steering Committee had initially envisioned. To help ensure this, Glass, the driving force behind the thematic approach, established the "theme corps," a subcommittee of six individuals assigned to oversee the revisions made during the second writing conference in order "to overcome the . . . lack of continuity of themes through the versions."[49] Problems however remained. John Moore, the primary author of the yellow version, was never all that happy with the collaborative approach to writing the textbooks. "I think BSCS has proved beyond a doubt," he wrote to Grobman in the fall of 1961, "that books cannot be written by committees." Since each writer came to the task with his or her own pet topics in mind, the end result was, not surprisingly, "a Moon, Mann, and Otto sort of compendium."[50] Adding to the difficulty was the thematic nature of the three versions themselves. Turnover among the lead writers and disagreement within the Steering Committee as to what approach or emphasis each—yellow, green, or blue—should showcase led to frustration and dissatisfaction among some of the biologists involved.[51] Perhaps the most unifying aspect of all the BSCS curricular materials, in the end, was the very lack of a unified vision of biology.

## The Return of Evolutionary Biology

Although the biologists struggled with thematic coherence in their writing, the curriculum materials they eventually produced did foreground a number of key ideas, if not all nine themes, that signaled BSCS's radical break with existing biology textbooks. The most controversial of these, and the one for which BSCS became so well known, was the open treatment of Darwinian evolution.[52] Evolution had long been regarded as the black sheep of the biology family of topics, at least since the Scopes trial in 1925, an event that prompted most textbook publishers to eliminate explicit references to evolution in order to avoid controversy.[53] With the entry of the country's top biologists into the field of pre-college science education, however, it was certain that Darwinian ideas would be welcomed back into the fold. Evolution as an organizing idea was too valuable to the project in which they were engaged to be allowed to lay fallow—it provided the key unifying framework that served to define the science of biology and, at the same time, provided perhaps the greatest test of the public's willingness to embrace the scientific rationalism of the postwar world.

By the 1950s, evolution had emerged within the biological community as a theoretical structure that many believed would finally reverse the fragmentation of the life sciences. Biologists by this time were hailing what came to be called the "modern synthesis," which was the intellectual union of evolutionary biology, previously viewed as speculative and lacking scientific rigor, with mathematically-based population genetics. This new quantitative foundation opened the door to experimental exploration of the mechanism of evolutionary change. Leading biologists such as Theodosius Dobzhansky, G. Ledyard Stebbins, Ernst Mayr, and George Gaylord Simpson invoked this more rigorous framework as a unifying element that provided explanatory power to disparate fields such as taxonomy and systematics, paleontology, and plant biology. With the arrival of the modern synthesis, many biologists viewed evolution as a disciplinary field on par with the mechanistic and materialistic disciplines of physics and chemistry. Biology could now be considered a science in its own right, yet—because of the progressive, emergent properties life held—still distinct from the physical sciences.[54]

This "maturation of biology and the emergence of evolutionary biology," in the words of historian V. B. Smocovitis, coincided with the centenary celebrations of the publication of Darwin's *Origin of Species* in 1959. A conference at the University of Chicago marking the event included notable speakers (such as Julian Huxley, noted biologist and former director-general of UNESCO), roundtable discussions, and even the performance of a musical play based on Darwin's life.[55] Also in attendance were a select group of high school biology teachers as part of a program to disseminate modern biological knowledge to the schools. This portion of the conference resulted in the production of a pamphlet that contained a summary of the issues raised with respect to teaching evolution and an article by Huxley entitled, "Evolution in the High School Curriculum."[56]

Members of the BSCS Steering Committee were active participants in the celebratory spirit of the Darwinian centenary and the re-emergence of evolutionary biology more generally. Arnold Grobman was a founding member of the Society for the Study of Evolution established in 1946.[57] Other BSCS supporters who extolled the unifying power of evolutionary theory included Stebbins, Simpson, and Nobel laureate Hermann J. Muller, who joined the project in 1961. Thus it is not surprising that at the first Steering Committee meeting, when the original nine themes were laid down, evolution was included without much comment. It would have been impossible for those present in the year of the *Origin's* centenary to imagine biology without it. Indeed, as a topic it was actually included twice, once as one of the key themes and again as one of the major divisions of biological subject matter.[58]

As important as evolution was for providing a unified view of the fragmented subdisciplines of biology for high school students, equally important was its prominent place at the point in the scientific community's battle against the various manifestations of postwar irrationalism. Of all the topics in science, evolution embodied, more than any other, the fruits of free inquiry. It exemplified the deep self-understanding of which humankind was capable if freed from the oppression of blind ideology, such as that found in the Soviet Union, and the anti-intellectualism of the masses characteristic of American society.[59]

The fundamental incompatibility of political ideology and scientific inquiry, as we have seen, was pointed out repeatedly by the American scientific elite in the late 1940s and 1950s. Evolutionary biology, particularly that grounded in molecular genetics, was the leading example of just how destructive political interference in scientific research could be. This was the point that Muller, a tireless apostle of modern evolutionary theory, came back to again and again in his popular writing. In a speech before an Indianapolis meeting of science teachers in 1958 entitled "One Hundred Years Without Darwinism Are Enough," he described the advantage America had over the Soviets in science because of its free society and democratic form of government. The Russians, he said, "take it for granted that politicians may dictate to scientists what the true principles of science are, that they may stifle opposing views, and that they may even prescribe in what direction research should be pursued." The unquestionable result of such policies was the "loss of the very soul of science." This was the fate of Soviet genetics, Muller explained (still stinging from the communist purges in that country), "a field basic to the understanding of biological evolution and of everything that has resulted therefrom."[60]

At home, the enemies of rationality were individuals cloaked not in red, but in the vestments of the growing evangelical religious groups of the mid-1950s and those who would bend to their will to avoid controversy.[61] The BSCS biologists felt that the schools and textbook publishers had kow-towed to this vocal minority long enough. The religious opposition to the teaching of evolution, one speaker lectured to the teachers gathered at the Darwin Centenary, "is firmly rooted in an essentially antiscientific approach to the natural world and is supported by law, by school boards whose members agree with the opposition, and by school administrators who dare not permit the use of controversial material in their schools."[62] This claim seems to be borne out by the reports of biology teachers in 1942 that indicated fewer than half treated the subject in their classrooms, many for fear of the controversy it might engender. From the 1920s through the early 1960s, nearly all textbooks sought to camouflage any discussion of the topic.[63] Grobman noted in 1961 that eight of the ten general biology textbooks in the BSCS offices did not list the word "evolution" in the index.[64]

The upshot of all this was that, in the words of a typically aghast Muller, the Darwinian theory "has not yet come to form an integral, fundamental part of the average American's way of thinking and outlook on life."[65] With some overstatement, he preached to the teachers assembled in Indiana that "we have no right to starve the masses of our people intellectually and emotionally because of the objections of the uninformed than we have a right to allow people to keep their children from being vaccinated and thus to endanger the whole community physically."[66] Emboldened by the nationwide calls for revitalizing science and hoping to ride on the "age of science" coattails of the physicists, the BSCS Steering Committee pressed forward in its efforts to present evolution as the central organizing theme of biology without apology. It was simply a decision that "we are not going to back down on," Glass insisted.[67]

Although there was definite consensus about the importance of including evolution in the curriculum, there was some disagreement regarding how best to present it in the curricular materials.[68] At the fourth Steering Committee meeting in February 1961, Muller expressed his dissatisfaction with the group's early efforts. Specifically, he was "disappointed to find the evolutionary thread missing in the versions appearing so far."

He felt that evolution had, perhaps, become an afterthought when instead "it should come at the beginning, all the way through and again at the end." Muller had never completely bought into the thematic plan Glass laid out at the first meeting. He went on to explain that "one could certainly thread the themes through now," but there really were not "nine separate themes," in biology. For Muller, evolution was "the trunk: the bottom, the top and everything," and the goal of any biology course should be to convey this to students.[69]

Part of Muller's argument to the Committee for a strong focus on evolution centered on the inherent power he believed the subject had to combat ideological oppression. For Muller, the realm of scientific practice was one of the most important battlegrounds in the Cold War with the Soviets. "More than half the world is under the influence of people behind the iron curtain. . . . It is up to us," he implored his colleagues, "to put up a strong front of the kind of biology we now think of." By presenting the truth of Darwinian evolution, he stated, "we can give ourselves and the world a logical science and the Russians will recognize this. The other side of the world doesn't recognize competition and the continuity of genetic material. To get this across would be our great contribution."[70] In the international rivalry between American science and Soviet science, Muller felt that it would be "a grievous step backwards if the Soviet system, with its authoritarianism and its perversion of biological as well as social progress, were to win out in the struggle for the minds, hearts and bodies of men."[71]

Muller's comments had a significant impact on the other biologists at the meeting. The Committee moved immediately to reconsider the emphasis it had been giving evolution up to that point. Glass, for one, was in almost perfect agreement with the sentiments expressed. He raised the possibility of producing a fourth textbook version—one devoted exclusively to the treatment of evolution closely tied to molecular biology and genetics, an approach that would capture the essence of the modern synthesis. Muller, of course, supported such a move. Others, however, argued that the production of a strongly evolutionary version might relieve the pressure to include evolution in the remaining versions, and as a result "many schools would not be exposed to this viewpoint at all." The Committee concluded that it would be best to limit the project to the three versions agreed upon, but to revise one of them to conform to the evolutionary framework Muller favored. In the end, all three textbooks reflected the importance of evolutionary theory as the unifying force in biology.

In contrast to past textbook treatments of the subject, which either portrayed Darwin's assertions as merely theoretical speculation, or simply stated the components of the theory dogmatically and then moved on, the

BSCS texts sought to convey both the richness and explanatory power of Darwin's ideas.[72] All the versions provided a discussion of the logical structure of the theory of evolution by natural selection with a detailed look at the evidence on which it was based. To varying degrees, the textbooks realized BSCS consultant William Mayer's appraisal of evolution as a subject that "offers one of the greatest opportunities to present the development of a scientific idea we find anywhere in biology."[73] The result was a convincing narrative that laid all the facts plainly before the student and, as textbooks do, drew for him or her the only reasonable conclusion that could be drawn—that life on earth had not been specially created, that it had instead evolved from simpler to more complex forms through the process of natural selection. The only choice left for the student to make was between, as G. G. Simpson phrased it in one of his popular articles, "simple rational acceptance or superstitious rejection."[74]

The early returns from the first-year testing of the curriculum materials seemed to confirm the Steering Committee's sense that the broader public was open to sound scientific knowledge in biology. Out of 105 test classrooms only one reported any objections to the teaching of evolution, and in that case the students asked only to be excused from the lessons. No one suggested that the offending sections be censored. "It seems to me," Grobman surmised from this report, that "insofar as the teaching of evolution in public high schools is concerned, the American people are far more rational than the commercial textbook houses believe." He also noted parenthetically the favorable influence of the recently released motion picture *Inherit the Wind,* which "probably contributed to a sympathy for the teaching of evolution by many persons otherwise apathetic."[75] Though not primarily concerned with the advancement of modern biology in the schools, the authors of *Inherit the Wind* advanced a favorable, if oversimplified, view of science as supremely rational and marching forward against the enemies of enlightenment—a view that gained a wide audience in the popular culture. As indicated by Grobman, the BSCS biologists, certainly opposed to irrationalism of any sort (ideological or religious), were not going to complain. Whatever might aid in the dissemination of the evolutionary point of view and advance the progress of science was a welcome ally to the cause.[76]

## Foundations of Progress

In addition to the explicit themes highlighted in the BSCS materials, of which evolution was one of the most prominent, there was an important implicit theme that helped define in more subtle ways the character of the new biology curriculum: the idea of biologically-directed human progress. In contrast to the physicists at MIT, who relentlessly cleansed their materials of all references to applied science, the biologists agreed at the outset to do just the opposite. Though committed to the belief that citizens must develop "an understanding of the very real distinction between pure science and that 'research and development' which takes most of our federal dollars," they, nevertheless, also insisted that "any textbooks or pamphlets to be published by the BSCS should include . . . everyday applications."[77] In this sense, the new biology textbooks seemed to have more in common with existing textbooks than with the newer NSF-sponsored materials. The difference, however, was that BSCS placed its emphasis not on the intersection of biology with the personal or social needs of the student, as one might find in the discredited life-adjustment education program, but rather on the intersection of the biological sciences with broader social issues of national interest. Indeed, there was a feeling among those involved that in the absence of sound biological understanding, the future of civilization might well be in jeopardy.[78]

The reason for the biologists' relative focus on the applied aspects of their work can be attributed partly to the nature of their intended audience. John Moore argued that "a high school biology course should be practical in the sense that it should provide the information that we expect educated citizens to possess," while at the college level more time would be devoted to the theoretical and abstract ideas of the discipline.[79] But another reason for the more practical focus can be traced to the status disparity that existed between biology and the physical sciences throughout the postwar period. The physicists, having already demonstrated the value of their disciplinary skills to the nation's interests, were, as a result, thoroughly immersed in the "big science" research enterprises favored by the various national security agencies, and the inevitable space race brewing with the Soviet Union promised more of the same. American biologists, for the most part, were left on the outside looking in, particularly with respect to government allocation of research funds. The value of their expertise to the national welfare had yet to be sufficiently recognized; the argument for the importance of biology remained to be made.

One such argument, strained as it was, could be found in AIBS director Hiden Cox's letter to the president's science advisor James Killian calling for the inclusion of biologists on government policy-making boards. In response to the public clamor and speculation about the possibilities of humans following the satellites into space, Cox insisted that "the problem of man in space is a biological one and not a simple one." Noting the obvious fact that "a segment of man's earthly environment must be carried with him into space," he attempted to make the case that "this can only be achieved after an exhaustive series of physiological, ecological, and morphological experiments have been performed by biologists commencing with simple experiments using simple forms of life, progressing to more sophisticated experiments on more complex organisms."[80] Even to the casual observer, Cox's proposed research program would have been seen as overly cautious.

More convincing arguments for biology's relevance were made regarding problems closer to home. The most succinct statement of these was contained in the 1956 AAAS report on the social problems of science. The report authors cited the rapid advancement of science as the very cause of the most pressing social problems of the day. These included hazards such as "the dangers to life from widely disseminated radiation, the burden of man-made chemicals, fumes, and smogs of unknown biological effect which we now absorb, large-scale deterioration of our natural resources and the potential of totally destructive war."[81] Such dangers required that the public "be alerted to humanity's dire need of future research and application of biology," as one biologist warned.[82] This type of argument reflecting immediate national interests made its way into the public statements of the BSCS program. Grobman, for example, claimed that a proper education in biology was necessary to eliminate among other things, "the regrettable tensions and misunderstandings between race groups, and . . . the conflicts and cold wars traceable to inefficient and wasteful uses of natural resources by burgeoning human populations."[83] There was certainly a sense among the biologists that both science and society were at a crossroads. Glass noted to himself at the third Steering Committee meeting in the winter of 1960 that "society is in turmoil. Education is in ferment. The growth of science and technology make imperative vast readjustments in social outlook and educational methods."[84] The wise application of science, from the BSCS perspective, seemed to be the surest road to salvation.

Each of the three BSCS textbooks included analyses of pressing social problems from a biological perspective. The green version, written primarily by Marston Bates, for instance, placed humankind squarely within the web of nature. The overriding point Bates sought to convey was that, whatever man's conceits, there was no escaping the fact that "the economy of nature and the ecology of man are inseparable and attempts to separate them are more than misleading, they are dangerous."[85] He began by introducing concepts such as the biosphere, populations, communities, and ecosystems in the early chapters of the text, elaborating in particular the scientific study of populations. He described the eighteenth-century ideas

of Malthus and went on to outline the various factors that influence population growth, using flour weevils, rats, and snowshoe hares as examples.
After surveying the biomes of the world and dipping into cellular biology,
physiology, genetics, and evolution in the middle chapters, he returned explicitly to a discussion of "the human animal." Here Bates addressed the
primary biological problems of modern man, specifically those related to
the population factors discussed earlier—health and disease, agricultural
production, and the depletion of natural resources. All brought the student
face to face with the problem of uncontrolled human population growth,
which Bates likened to "the growth of a cancerous tissue within an organism." He concluded with a warning that "there is a frightening possibility that man, with his apparently limitless increase in numbers and his
increasing power to destroy the rest of nature, may multiply his way to destruction," unless of course he recognized the biological consequences of
his current way of life.[86] Thus, the population bomb joined the atomic
bomb as source of national anxiety.

All the versions contained biological treatments of human racial groups,
no doubt a topic uppermost in the minds of Steering Committee members during the exploding civil rights movement in the early 1960s.[87] At
the eighth Steering Committee meeting, following the freedom rides in
the summer of 1961 and the riots at Ole Miss in the fall of 1962, Theodore
Puck of the University of Colorado-Denver described some unsolicited
pamphlets his department had received "composed of highly scientific
sounding but inaccurate analyses of problems of human race." His main
concern was whether the new textbooks were "going to give students
enough ammunition to handle this sort of situation." Glass replied that
there was "a chapter in the Yellow Version on human races which he felt
was beautifully done," along with a BSCS pamphlet on the subject in the
works.[88] The text to which Glass referred made clear to the student that
"despite the fact that we can divide *Homo sapiens* into races on the basis of
percentage differences of many inheritable traits, the different members of
the human species are still much more alike than they are different." The
text went on to state unambiguously that "by biological criteria, all men
are of one species," and that any differences between human groups are almost exclusively the result of "culture and tradition." These statements, the
biologists hoped, would address common biological misunderstandings
about the nature of man and, thus, reduce racist tendencies in society.[89]

The blue version, with its molecular approach, provided a detailed explanation of the biological effects of radiation on both the body and reproductive cells of mammals. With respect to humans, the effects on the
reproductive cells were deemed "the more important ones," particularly

from the "standpoints of genetics and evolution." Changes in these cells, the text noted, "may affect all future generations."[90] The damaging effects of radiation was a topic of particular interest to Glass, who embraced efforts to inform the public about the dangers of fallout that resulted from the increasingly frequent atmospheric tests of nuclear weapons, despite statements from the Atomic Energy Commission that minimized such risks.[91] Some BSCS members questioned whether the inclusion of material on the "effect of radioactivity on life" should be considered as fundamental to biological science as, say, "the patterns of information transfer from the nucleus to the cell."[92] Glass, however, insisted, and successfully added both a monograph about atomic energy and a laboratory block on radiation biology to the BSCS project list. However such materials might detract from the discipline of biology in its purest sense, they, nonetheless, served to tie the biological sciences to one of the most pressing issues of the new scientific age. Disciplinary integrity, in this case, took a back seat to social relevance.[93]

In all these instances, the directors of the BSCS project hoped to demonstrate not just the relevance of biology to society, but, more important in their view, the crucial need for biological understanding in the management of public affairs. Unlike physics, the subject of biology, especially as it related to humans, could never be treated in isolation from the broader societal issues and constraints it inevitably ran up against. It was obvious to all concerned that, as Bates clearly realized, "the problem of man's place in nature . . . is the problem of the relations between man's developing culture and other aspects of the biosphere."[94] One certainly could pretend that man did not exist, as Bates observed a good number of biology textbooks did, but there would then be little basis on which to argue for the importance of biologists to American interests. An appreciation of the biological foundations of current social problems, the BSCS team believed, would provide the necessary perspective through which rational solutions, in harmony with the laws of nature, might be sought. It was only through such a biological approach that progress could be made in resolving the longstanding social problems of man, such as racism and overpopulation, as well as those that physical science had more recently opened up for the world as it ran ahead of the rest of the scientific pack.

The application of biological expertise to these societal threats was only the most immediate part of what the biologists saw as the much grander role of their science in the progress of human civilization. The resurgence of the Darwinian point of view, within both the scientific community and popular culture, created an intellectual climate in which many biologists saw some hope for pulling civilization out from under the ever-multiplying social, political, and scientific problems of the Cold War era. During this time, Smocovitis notes, America "resonated with evolutionary themes. . . . Technological progress and evolutionary progress, inextricably linked, were to hold the key to the future of the most intelligent and unique species on earth: 'modern man.'"[95] Central to this idea of progress, as Glass had explained to the teachers in Colorado, was man's ability to contemplate his own actions with an eye to directing and even accelerating the natural evolutionary process of the human species—a process that included the dynamic nature of human culture and ideas. Whatever the future might hold it was ultimately grounded in the biology of man. As Simpson explained, human culture itself was not separate from nature, but was itself a "biological adaptation." This culture, in turn, "evolves not distinct from and not in replacement of but in addition to biological evolution, which is also continued."[96]

The topic of cultural evolution received a good deal of play throughout the BSCS program. Social problems, human progress, and broader notions of culture were all intermingled in the biological treatment of humans. Each of the BSCS textbooks described in detail the organic evolution of man from *zinjanthropus* (a species that lived approximately 1,750,000 years ago) to modern humans.[97] These discussions of biological evolution graded, almost imperceptibly, into descriptions of the progressive cultural phases of the human species. Milestones such as the development of tools, the domestication of plants and animals, and the subsequent impact of human society on the environment were marked as students were invited to consider what the future might hold for humanity in an age when, in the words of the AAAS report, "the forces and processes now coming under human control are beginning to match in size and intensity those of nature itself."[98]

The yellow version contained the most explicit discussion of cultural evolution as a biological process. It was, in fact, described as a "new kind of evolution" that man had discovered, "which is independent of the physical changes in his body." This section of the text introduced students to some of the key characteristics of this newer type of evolution: the concept of *cultural inheritance* as an analog to genetic inheritance, the power of cultural inheritance to enable man to recreate his environment to meet his needs, the advantages of learned cooperation as a cultural trait, and finally "the ability of some men and women to look ahead and foresee a better world" as a natural product of this process of cultural development.[99]

Despite the challenges faced by humans in the coming decades, Muller for one had "enough confidence in the ultimate ability of human beings to see things straight, and in their sense of responsibility, to believe that once they are aware of the facts of genetics and evolution they will be willing to repattern their behavior in conformity with the long-term view." The biological and cultural potential seemed unlimited given a rational approach to human problems. "We know we can influence [evolutionary] processes and we have made good starts in the alteration . . . of such organisms as fruit flies," Muller explained. "Undoubtedly what [we] will know and be able to do along biological lines only a few generations hence would seem like a science-fiction dream to most of us today." Since primitive times man has "continued to remake himself culturally," but the enlightened combination of directed biological and cultural evolution was "far more effective than either course alone."[100]

The BSCS biologists saw their work, interestingly enough, not just as a means to generate greater understanding of modern biology, but in so doing to facilitate the very progress of humankind that they described in their textbooks. In one of the early grant proposals to NSF, the biologists cast their project unambiguously as a way to make a real contribution "to an understanding of society and to the expected direction of its evolution."[101] The place of science and science education in the evolutionary process seemed clear. The authors of the yellow version explained that "education in science would appear to be an essential, central matter. It seems obvious that it is not safe for apes to play with atoms. Neither can men who have relinquished their birthright of scientific knowledge expect to rule themselves." They insisted to the student: "You must . . . see the science of biology as a part of man's cultural evolution." And in grander terms they implored their readers "to appreciate the nature of this human conquest which . . . makes man like the ancient gods in his power to control nature and to work 'miracles.'"[102] In a draft public statement regarding the role of BSCS, one writer, using a rather industrial metaphor, noted that "the rate at which science is . . . modifying itself and our lives is outstripping the machinery of communication which formally sufficed to bring us understanding of those changes. It is the hope of the Biological Sciences Curriculum Study to so complement this machinery that its function will be sustained."[103] Thus the BSCS curriculum itself was to become, in a small way, part of the mainspring that would help drive the future evolution of humankind.

### Teaching Science as Inquiry

As important as the topics of human progress and evolution were for establishing the place of biology in American society, the proper appreciation of these ideas could come only through an understanding of the nature of

biological science as a means through which humans are able to comprehend the living world. Little would be accomplished, biologists contended,
if students merely learned by rote the steps in Darwin's model of natural selection or the many ways biology had contributed to human welfare. The
common thread uniting these ideas was the intellectual process by which
the progressive power of organic and cultural evolution had been discovered, the process that would provide solutions to the problems of the present and future. The deliberate construction of "science as enquiry" (as it
was listed among the nine BSCS themes) was seen as "one of the most radical departures of the BSCS texts from conventional patterns."[104] This emphasis on scientific process bound biology to the fraternity of scientific
disciplines, both philosophically and politically. Here BSCS, having carved
out a niche apart from the physical sciences, rejoined the advancement project of science laid out by PSSC and the scientific elite.

The importance of teaching about scientific inquiry was accepted by
the Steering Committee without question. Common among scientists of
the time were statements such as Muller's that "the trouble is not that there
is too much science but too much short-sighted application of it, too little dissemination of its deeper meanings, and too little appreciation of the
need for proceeding by its method of free inquiry . . . in every sphere of
existence and of thought."[105] Paulson, the NSF liaison to BSCS, had earlier posed the question, "Is it asking too much of one or a few science
courses to expect them to give individuals a real sense of the power of
these methods of inquiry, and of their value and importance, when from
cradle on there is so much other teaching that is frankly and deliberately
authoritarian?"[106] The difficulty was, as always, in developing materials that
conveyed the appropriate conception of this process to students. It was a
problem that, as Bates commented at one of the early BSCS meetings, "no
textbook has licked" yet.[107]

The biologists, as the physicists before them, were clear at least about
what science was not. Glass restated the disagreement scientists had with the
common view: "Scientists deplore the popular image of science as a benev-
olent genie who will provide any gift the Master of the Lamp may demand,
or the popular conception of the scientific method as a sort of 'intellectual
machine that inevitably grinds out ultimate truth in a series of orderly, predictably sequential "steps," with complete accuracy and certainty.'"[108] After
the second year of the project got underway, John Moore volunteered to
put together a statement listing the primary goals of the new curriculum.
With respect to the nature of science, he wrote that science should be understood as "an open-ended (ever expanding) intellectual activity and what
is presently 'known' or believed is subject to 'change without notice'; that

the scientist in his work strives to be honest, exact, and (part of a community) devoted to the pursuit of truth; that his methods are increasingly exact and the procedures themselves are increasingly self-correcting."[109] Vague as it was, this characterization of scientific work captured the sense of the Committee. Intended to suffuse all the BSCS curriculum materials, this view of science as a community-based, self-regulating activity reflected the image of science advanced by the scientific elite throughout the 1950s, an image concerned with securing both autonomy and funding for science—goals the physicists and biologists shared equally.

BSCS's emphasis on inquiry and the nature of its manifestation in the various components of the new curriculum owed a great deal to the intellectual presence of Joseph Schwab. Schwab, who received his Ph.D. in genetics under Sewall Wright at the University of Chicago, made an academic career out of examining the central role of inquiry in the activities of scientists and the place of science in a liberal education. This he did in his capacity as an instructor in the University of Chicago's undergraduate general education program, a program similar in its educational goals to Conant's at Harvard.[110] The exponential advancement of science over the course of the twentieth century made it imperative, Schwab argued, that science teaching be redirected from its current presentation of science as a static body of knowledge—what he termed a "rhetoric of conclusions"—to an examination of the process by which that knowledge was generated. It was the revisionary character of scientific knowledge that was important for students to understand, that the conceptual structures of science were not ends in themselves, but should be viewed instead as "instruments of inquiry, pragmatic and constructive, and not as assertions about the nature of things."[111] Ironically, given the general disdain among scientists for Deweyan ideas, Schwab's writing on inquiry and education was heavily influenced by his reading of Dewey in the late 1940s. It was unlikely, though, that any scientists would have recognized this philosophical foundation of his work.[112]

In a series of essays, Schwab laid out the educational rationale for why a "radical overhaul" of the science curriculum along these lines was necessary.[113] The need for more scientists and engineers to staff the expanding national laboratories was certainly the easiest justification to offer given the technological race with the Soviet Union. Echoing the call for basic research made by Vannevar Bush some 15 years earlier in *Science—The Endless Frontier,* Schwab acknowledged that original scientific inquiry was increasingly necessary to advance the boundaries of knowledge from which technological applications were drawn. An inquiry approach in the classroom would thus provide the kind of intellectual excitement necessary to "permit the imaginative, the able manipulator of symbolic structures, the exploiter of novel and remote possibilities, to identify science as an area in which there is scope for his talents."[114] This argument for replenishing the stock of scientific personnel, however, was no more compelling for Schwab than it was for the physicists. His primary concern lay with the perception of science by the general public. But rather than insisting on the need for curriculum reform to promote a greater degree of public rationality, from which support for basic scientific work would, in turn, naturally emanate, one might say Schwab cut to the chase. He justified public understanding of science almost exclusively in terms of ensuring continued support of the U.S. scientific-research establishment.

This argument was directly informed by the historical and political circumstances scientists found themselves in at the center of the technological stand-off with the Soviets. Schwab described what had become commonplace by the mid-1950s, that in an expert society such as the United States "the specialist is no longer merely the source of information, and the lay officer the maker of policy. The specialist, of necessity, participates in policy making, and the lay officer can no longer, by himself, determine and correlate the relevant facts."[115] Consonant with the view of the expert society outlined in the President's Science Advisory Committee's statement on education, Schwab concluded that the scientific expert had become an essential figure in the governance of modern society. The scientists insisted, however, that the value of their expertise depended upon the pursuit of research unfettered by government interference and supported financially by the public.[116]

In "The Teaching of Science as Enquiry," Schwab's most comprehensive and perhaps most influential essay on the need for inquiry-based science education, he provided an example of the danger that traditional, content-centered curricula held for science. "Consider a student who has garnered the impression that science consists of inalterable truths," he invited the reader. "Five or ten years after graduation he discovers that many of the matters taught him are no longer taught. . . . They have become obsolete and been replaced by other formulations." The likely result, Schwab imagined, is that "the former student, now a voting member of the polity, can do no better than to doubt the soundness of his textbook and his teacher. In a great many cases, this doubt of teacher and textbook becomes a doubt of science itself. . . . [He] has no recourse but to fall into a dangerous relativism or cynicism . . . [concluding] that experts are untrustworthy, parading the doubtful as certain, or even that they promulgate particular views as true because it is to their own interest that people believe in them."[117] Thus a curriculum that presented science as a static body of knowledge, regardless of how accurate and up-to-date, Schwab believed, contributed to "a climate of opinion inimical to science."[118] With science now the "foundation of national power," what was essential and "perhaps most important" was the "need for a public which is aware of the conditions and character of scientific inquiry, which understands the anxieties and disappointments that attend it, and which is, therefore, prepared to give science the continuing support which it requires."[119] Bringing the public to this enlightened view—developing "a voting and supporting republic of non-scientists"—Schwab asserted, "constitutes the main burden of the secondary school."[120]

In the *Biology Teachers' Handbook,* the companion to the BSCS curriculum, Schwab described the ways in which inquiry was foregrounded in the various curricular materials. The textbooks themselves sought to emphasize the uncertainty of science at every turn, using phrases such as "we do not know," or "the evidence about this is contradictory," throughout. The books also worked to present ideas using what Schwab called a "narrative of inquiry," the objective of which was to trace the intellectual development of current scientific ideas, examining the experiments performed and the interpretations of data made along the way.[121]

Beyond the textbook, the BSCS developers experimented with less conventional instructional materials designed to place the student in situations where he or she would actively participate in the inquiry process. The BSCS film group, for example, developed a set of innovative "single topic films" as they were called. These were short films, only four to five minutes long, that presented students with visual data that they were asked to interpret. The film would be stopped while the class discussed the various interpretations and proposed experiments or observations that could be made to test some of the hypotheses suggested. The teacher would then resume the film, which would provide additional data that the students might need to take into account.[122] The most successful of these early single topic films was one entitled "Social Behavior in Chickens," which was designed to have students come up with the idea of peck order after careful observation of the social interactions among a group of individual birds.[123]

The lab block program provided a more direct avenue for students to appreciate the active process of inquiry. The goal of this program, the "brain-child" of Bentley Glass, was to focus exclusively on developing the more investigatory aspects of scientific work, relegating the illustrative function of the science laboratory—which often took the form of cookbook-type exercises—to other portions of the curriculum. "It is in the laboratory," Glass resolutely claimed, "that the work of science is done; that the difference between science and magic on the one hand, and science and authoritative teaching on the other, is clearly perceived at first hand."[124] Each block was to be developed by "a university scientist of standing, and one whose work [lay] in that particular field."[125] The intensive investigation of a single topic or phenomena, such as homeostasis or animal behavior, would involve "a series of activities from which the student can tie together answers and ideas which will enable him to develop for himself some of the fundamental principles of biology."[126] The student might begin by replicating a few of the classic experiments in the field, but would soon "arrive at the frontiers of knowledge, at least in regard to some particulars," Glass explained.[127]

Schwab's own contribution to the BSCS inquiry materials were the "Invitations to Enquiry," which were short written vignettes that brought before the student "small samples of the operation of enquiry, samples graded to his competence and knowledge" that served as the basis for class discussion.[128] Moore commented at one meeting that, in his estimation, the "Invitations" were "the most interesting things to come out of BSCS."[129] In operation, the teacher presented the background to the invitation in class, which usually included an incomplete data set along with other information. Students were then expected to volunteer their analysis of the situation. The teacher was to follow up with "diagnostic questions," which were to "help students see what [was] wrong with poorer answers," and then review "the logic that justifie[d] sound responses." The vignettes, of which there were 44 in all, were progressively more difficult, each focusing on different aspects of scientific inquiry, such as the consideration of experimental controls or the nature of serial causation.[130]

The proximate learning outcome of these various curricular components, in Schwab's words, was to "show students how knowledge arises from the interpretation of data, . . . that the interpretation of data . . . proceeds on the basis of concepts and assumptions that change as our knowledge grows," and that "because these principles and concepts change, knowledge changes too."[131] Ultimately, though, it was clear that public understanding of the revisionary nature of science—science as inquiry—had everything to do with the maintenance of the expert society that the United States had embraced following World War II.

During the summer of 1963, the green, yellow, and blue writing teams, along with the other various subcommittees, put the finishing touches on the assorted curriculum materials in preparation for the coming school year.[132] The commercial publication of the BSCS program came three years after PSSC. Following the physicists, the biologists had built into their curriculum a portrayal of science that was sensitive to the political and social concerns expressed by the physical science elite—primarily those regarding funding and autonomy. But while PSSC sought to shape public understanding of science as an intellectual activity divorced from its social applications (for fear of having the American public misunderstand the fundamental nature of scientific research), the biologists, in a bid to garner a more prestigious and influential place for biology in American society, sought through their curriculum to demonstrate the social relevance of the biological sciences. Scientifically grounded views of organic and cultural evolution, driven and directed by the universal methods of scientific inquiry, they believed, provided both hope for the future in dangerously uncertain times, as well as sound justification for the importance of the community of biologists to a modern technocratic society.

CHAPTER 7

# SCIENCE/EDUCATION TRANSFORMED

In May 1960, while PSSC was preparing its course for nationwide distribution and BSCS was deep into its writing phase, John Hersey's book *The Child Buyer* was published. Written in the form of hearings before a senate committee on education, welfare, and public morality, the novel described the fictional exploits of Wissey Jones, a vice president in charge of human procurement for the U.S. conglomerate United Lymphomilloid, following his arrival in the small New England town of Pequot. His purpose there was to secure the purchase of a local ten-year-old child prodigy. As the story unfolds, the reader learns that U. Lympho, as the company is known, has contracted with the federal government to conduct top-secret research crucial to the country's national security, and Jones is out to accumulate the raw material for the project. "I buy brains," he states under questioning, brains that have not yet "been spoiled by . . . what passes for education"; brains that will be accelerated and enhanced through a highly technical process developed at U. Lympho and administered through Hack Sawyer University, an institution of higher education that might one day provide a template for a new kind of learning.[1] The training process, Jones explained to the senate committee in the book, had the power to multiply the IQ of its specimens nearly tenfold, primarily by eliminating the emotional energy brilliant children often spent in the largely futile search "for meaning, for values, for the significance of life."[2]

Hersey's book was a scathing critique of the frenzied push for more math and science education that gripped the country following the launch of *Sputnik*. It was an effort to restore balance and perspective to a nation that in its near hysterical reaction to the Soviet threat ushered in the National Defense Education Act and pushed Hyman Rickover's 1959 book,

*Education and Freedom,* which called for the sorting of students by ability among other things, to the top of the bestseller list. From our present vantage point, *The Child Buyer,* though not directed specifically at either PSSC or BSCS, serves as a useful historical artifact that can be used to draw out some of the complexities of the events surrounding this scientist-led effort to remake science education. In Hersey's characterization of the state of schooling in 1960, one can find evidence of both the oversimplifications to which public debate often succumbs alongside the essence of a deeper truth about the nature of the educational reforms that were advocated by so many individuals and institutions.

The broad-brush treatment of the events in question is apparent at the outset. The characters of Hersey's book, from the hopelessly muddled local school principal to the pompous, intellectually-challenged senators, to the cold, calculating corporate executive, are bitingly accurate in many ways. Through them Hersey laid out the various public attitudes held by key interest groups of the time, attitudes that facilitated the entry of scientists into the field of education reform. The Cold-War-conflict mentality that he used to caricature these players had achieved a media saturation that spilled over onto nearly every one and every thing during this era, including the PSSC and BSCS projects—all were painted with the same reactionary brush. Though the scientists, as we have seen in earlier chapters, were clearly not engaged in the enterprise Hersey described in his novel, at the same time, they were neither engaged in the enterprise they thought they were—the dissemination of science in its purest form as part of the cultural heritage of humankind, science that would, in turn, shepherd society out of the darkness of the Cold War.

The nature of scientific research in the years during and after World War II had begun to shift as a result of the U.S. government's interest in using science to safeguard national security. It was a shift from a more descriptive science of an earlier time, the primary aim of which was understanding as an end in itself, to science as technique and control, science as a means to other ends. It was a trend that, in the wake of the public fascination with the technological marvels of World War II, the new curricula had been developed to counter. Yet, in the end, the transformation of the scientific enterprise in the United States was perhaps too encompassing, too profound. The power of corporate and state interests, as exemplified by Hersey's powerful U. Lympho, appeared to be too much for even the most enlightened scientists to overcome, and science education, one might say, was itself co-opted to serve the dominant social and political interests of the time.

## The Myth of Technical Training

Hersey's book, though largely overlooked today as a literary work, garnered a great deal of attention as a critical commentary on the emerging
paradigms of education when it appeared in the fall of 1960. Hersey was a
Pulitzer prize-winning author whose work deftly explored the various
cultural tensions and ironies inherent in the postwar world. His book, *Hiroshima,* a vivid account of the atomic explosion over that city told through
the lives of six survivors, had created a public sensation when it first came
out in the *New Yorker* in the summer of 1946. *The Child Buyer,* his eighth
novel, had strong sales, was offered as a Book-of-the-Month Club selection, and led the *Saturday Review* critics' list of the most stimulating books
published that fall. It represented a culmination of Hersey's decade-long
interest in American education during which time he had served on a local
school board, participated as a delegate to the White House Conference on
Education in 1955, and worked as a consultant to the Ford Foundation's
Fund for the Advancement of Education. Whether applauded or condemned, Hersey's satirical portrayal of the state of schooling during the
years of the post-*Sputnik* hysteria seemed to reflect the genuine public perception of what education in the United States had become—an all-out
effort to train large numbers of scientists and engineers.[3]

Widely reviewed in periodicals such as *Time* and *Newsweek, The Child
Buyer* received its most extended public discussion in the pages of the *New
Republic.* Editors there had arranged for the simultaneous publication of
five reviews from individuals of various professional backgrounds, including a school superintendent, a poet, and a literary critic, among others.[4]
Novelist Margaret Halsey led off praising Hersey for his "starkly coura-
geous" look at "what we have let ourselves become." The book captured
for her all that was wrong with Cold War America—"our cowardice in
not defending McCarthy's victims, our refusal to take true responsibility
for our children, our concurrence in the exploitation of human emotion
for the sake of larger sales."[5] Most praised Hersey's commentary on the
incredible poverty of the country's educational vision, both prior to and
after *Sputnik.* The book roundly took to task the vacuity of the professional education establishment, a group he cast as one that unflinchingly
lumped the intellectually gifted with the physically handicapped and
mentally retarded as simply another deviant from the norm. The brunt of
his criticism, though, was reserved for what he saw taking its place—a
government-backed system designed to identify and develop by whatever
means the most intellectually capable individuals in the name of national
security. It was this "frightening demonstration of the possibility of a

monstrous misuse of education" that reviewers seized upon as the book's most important contribution. Nearly all condemned the idea of any "'educational' scheme which would transmute into a pure intellectual mechanism the free will and ranging spirit of the liberal mind."[6] The exception was B. F. Skinner, the Harvard psychologist and inventor of the teaching machine, an automated device built to facilitate self-paced student learning, which along with the use of film, television, and other instructional apparatus made up a significant portion of the war's contribution to education. Skinner, who the *New Republic* editors noted in a sidebar could be the villain of Hersey's novel, not surprisingly, objected to the unfair caricature of the new programs being developed. He chided Hersey for his simplistic story and unwillingness, or inability, to offer a more desirable alternative. In the more literate media outlets such as the *New Republic,* such a stance was rare.

What is most interesting is that the public debate over the direction of education reform seemed to center on the rather straightforward question of whether one was for it or against it. Few made any effort to challenge the singular characterization of the reforms being held up for scrutiny in these public forums. At a fundamental level, Hersey had given voice (and certainly further contributed) to the prevailing perceptions of the time that pre-college education was increasingly being driven by federal programs designed to increase scientific manpower resources—that is, education devoted to the promotion of science for future scientists. The only issue openly discussed was whether such an approach was insidious and dehumanizing, as some critics charged, or merely the most efficient way of satisfying real national needs, as Skinner and others insisted.

The most vocal, as is often the case, tended to be critics. Many derided the narrow science emphasis of the suggested reforms and called instead for a broad program of federal aid to update facilities, recruit teachers, and develop material in all content areas. Pulitzer prize-winning poet Karl Shapiro lamented what he saw as the government's antagonistic stance toward the humanities, arguing that the country had gone too far "in adopting scientific education." Others viewed the science-laden proposals as ill-advised attempts to emulate the Russian totalitarian system of schooling—a move that they claimed would ultimately crush cherished democratic freedoms.[7] The new programs were certainly not what Illinois historian Arthur Bestor had in mind when he called for a permanent scientific and scholarly commission on secondary education back in 1952. In a personal note to Hersey, in fact, Bestor expressed his admiration of *The Child Buyer,* lamenting the "horrors of a possible tomorrow" that Hersey had so vividly depicted.[8]

That the reforms spearheaded by NSF and other government agencies such as the Office of Education were viewed this way by the casual observer should come as no surprise. As we have seen, the motivation for federal involvement in reform came from the Soviet manpower challenge described in Nicholas DeWitt's book that so impressed Congressman Albert Thomas and other federal officials early in 1956. And following *Sputnik,* one could hardly talk about education without in the same breath mentioning national defense. Throughout this period, the national media was filled with images of Soviet proficiency in science and mathematics— images pushed by agencies such as the Office of Defense Mobilization as well as magazine and newspaper editors across the country. The *Life* magazine series on the crisis in education publicly demanded more rigorous scientific training to meet the Soviet threat, and Rickover's book, the most widely publicized critique in the post-*Sputnik* era, echoed that call. At the federal policy level, the National Defense Education Act of 1958 contributed significantly to molding public perceptions of the nature of the reforms being undertaken. Indeed the framing of the problem in defense and manpower terms was essential for any federal action to occur. Furthermore, it did not help matters that NSF officials often allowed arguments for manpower and defense readiness to pass without comment or clarification during appropriation hearings when funding decisions were on the line.[9]

What is surprising is the extent to which this interpretation of events during the tumultuous years of the late 1950s and early 1960s has endured largely unaltered to the present. Two points in particular seem to have been accepted uncritically by historians of this era: First, that the reforms were initiated for the purpose of increasing the number of scientists and engineers, and, second (closely related to the first) that the new curricular materials developed were narrowly technical, focused only on the hard-science content of the disciplines in question. Some writers have described what transpired as a "turning back from 'life adjustment' to the earlier goal of mental training," while others have portrayed the goals of the reformers as an effort to "align the curriculum of the public schools with the needs of American foreign policy during the Cold War for the purpose of educating more scientists and engineers."[10] The most recent historical account of the movement labeled all the "textbooks, laboratory manuals, equipment, and teacher-training programs these scholars devised . . . [as] oriented toward 'science for scientists.'"[11] As we have seen, these are half-truths at best.

What was lost amid the cries for more scientists in the 1950s and continues to be missing in our present-day understanding are the voices of the scientists themselves. From the perspective of Zacharias, Glass, Grobman, or Friedman, the suggestion that they intended to bolster their own numbers or sought to foist rigid disciplinary content on the nation's schoolchildren would have been seen as a profound misunderstanding of their work. Though the nature of the curriculum products generated did not preclude students from exploring careers in science, they were specifically designed to foster an understanding of science among the majority destined for nonscientific occupations. As Zacharias insisted, it was the "American people, and not just a few persons destined to become scientists" that needed to understand what science was all about, those who would become "the lawyers, businessmen, statesmen, and other professionals."[12] The same was true for BSCS, which was why the members of the Steering Committee decided to keep their course at the tenth-grade level, where it would help "as many high school students as possible be scientifically literate."[13]

As for the substance of the new science education, it was intended to be almost the exact opposite of technical training. There was intellectual rigor to be sure, but the new curricula were clearly conceived in the spirit of a broad liberal education. Amid the clamor for more raw training to match the Soviet manpower output, leading scientists repeatedly sought to get this message out through the various commission reports and national policy statements they generated. There was, of course, the white paper, *Education for the Age of Science,* put out by Zacharias and the President's Science Advisory Committee. But there was also the Rockefeller report, *The Pursuit of Excellence,* and, in the winter of 1958, a conference at Yale sponsored by the President's Committee on Scientists and Engineers, where the participants (including Zacharias, Margaret Mead, and Alan Waterman) affirmed that the goal of education was not to produce more scientists, but rather a "change in the environment in which . . . scientists work."[14] Finally, there was the statement on science from the President's Commission on National Goals, in which Warren Weaver argued for closing the gap that has opened up between scientists and "the rest of our culture." We must "turn our thought . . . away from gadgets, rockets, antibiotics, [and] plastics, . . . and recognize that science is in fact a noble intellectual and artistic pursuit."[15]

This emphasis on the humanistic aspects of science, the intellectual artistry of scientific work distinct from its dirty applications whether military or commercial, was, as demonstrated in the previous chapters, the essence of the education reforms of PSSC and BSCS. In his cultural advocacy of science, in fact, Weaver cited PSSC by name as an example of some of the "superb" projects already making headway with the American people.[16] This point has been well made. Yet the very persistence with which the scientific community advocated this pure, humanistic vision of science suggests the presence of some deeper, more profound problem. Something more seems to have been at stake than mere public misunderstanding. The scientists were seeking to clarify the nature of their work for the public to be sure, but they were also, perhaps, attempting to redefine it for themselves; for as we will see, the massive amount of government patronage and the demands of national security had begun to exact a heavy toll on the pristine idea of inquiry the scientists held so dear.

### Blurring the Lines

Though *The Child Buyer* failed to capture the intentions of the scientific community in its pursuit of education reform, Hersey's book hit the mark in another important respect. In his description of the rapacious, coordinated efforts of United Lymphomilloid and its academic partner Hack Sawyer University, he had identified a potent new social force that had arrived on the American scene in the 1950s—the military-industrial complex made famous in Eisenhower's farewell address to the nation delivered only months after the book's publication. This network of interests, which some historians have updated to the "military-industrial-academic" complex, shared a commitment to a set of goals and values that increasingly permeated all of science and science education by the mid-1960s, the most overriding of which was the desire to perfect techniques of manipulation and control of natural and social systems.[17] This goal when combined with the humanistic, cultural vision of science and science education resulted in significant tensions within the scientific community that are worth exploring in greater detail.

The various analytical techniques and organizational models that emerged from the war and were subsequently drawn upon in the summer studies made famous by Zacharias were viewed by many in the scientific community as incidental byproducts of their expertise in science. The sophisticated analytical skills needed for fundamental research had certainly proven themselves useful when applied to the practical matters of weapons development and defense strategy. This was not to obscure the fact, however, that such skills existed first and foremost for the pursuit of fundamental research, for the exploration of pure theoretical and phenomenological questions in the most esoteric realms. This was indeed the distinction the scientists were working so hard to preserve. In the field of education, this was made quite clear. The vigor with which the editors of the AAAS flagship publication *Science* applauded Hersey's scathing commentary on education reform provides an unambiguous indication of where the scientists saw themselves in this regard. *The Child Buyer,* they noted, should serve "as an antidote to that poison in our thinking which treats scientific creativity—and persons—as components in a system geared to larger ends."[18] Despite their reliance on such wartime techniques, scientists perceived themselves as collectively committed to the pursuit of far more noble, apolitical ends. Hersey's editor was quite pleased that the scientists were in their corner, commenting to him that "the fact that they have responded . . . so favorably is a hopeful sign for the country, apparently the scientists are no more willing to be mobilized by political fanatics than are the humanists."[19]

The individuals at PSSC and BSCS would indeed have publicly objected to any characterization of their efforts as manpower mobilization. Throughout their work they maintained a sharp distinction between the medium and the message. The coordinated use of various technologies elaborated at Woods Hole in 1959 (instructional films, innovative laboratory materials, textbooks, and so on) to teach science did nothing to preclude a decidedly nontechnical, humanistic learning outcome, science as cultural pursuit, at least as they conceived it. These techniques were seen simply as the best means of helping students and ultimately the public at large understand the commonsense foundations of scientific practice. The fact that Zacharias spoke to students about the pressure of light in thousands of classrooms via the medium of film, for example, in no way seemed to undermine his careful, stepwise demonstration of what it meant to rationally analyze a seemingly complex physical phenomenon. The physics itself and its patterns of reasoning remained pure. But despite the desire of the scientists to maintain this distinction, the line between ends and means increasingly blurred in the years after the war, both in their education reform work and their own scientific research.

The specific goal of raising public levels of reason and rationality through the curriculum using scientific thinking as the exemplar provides one case in point. Joseph Schwab of BSCS worried that, given the dire social and political circumstances of the time, without widespread attainment of some basic critical-thinking skills, the public would become prey to wholesale external manipulation. Though we will continue to "have the illusion of full self-determination," he wrote, we will end up "willing victims of our own subjugation. . . . Techniques of 'consensus engineering' are already in use by self-selected elites and taught by them to chosen disciples," he warned. "The mastery of these techniques constitutes the new demagoguery. These are its new methods."[20] Ironically, the importance of fostering the public's ability to reason intelligently (along with disseminating the appropriate images of science) required that compromises be made. With the public climate for science so seemingly unfavorable, the new techniques of mass communication and consensus engineering proved too tempting to overlook. Ultimately all the talk about reason and rationality, however sincere, ended up as hollow rhetoric. Expediency won out over principle.

Scientists and government officials had only to look back as far as Project Troy, the State Department counter-propaganda study held at MIT in 1950, to find the most productive way to frame the problem of getting the new curriculum projects into classrooms. The carryover in personnel from Troy to PSSC ensured that information dissemination techniques perfected for use overseas would find a productive outlet at home as well. At PSSC, Zacharias and the Steering Committee began to map out a dissemination strategy for their new course. Their self-described task was to "find mechanisms for completing the production of materials . . . and to arrange smooth mechanisms for indoctrination of teachers, distribution of materials, and evaluation of their effectiveness."[21] To reach as many physics teachers as possible, the physicists, in typical systems engineering fashion, sought to tap the potential of the rapidly growing NSF summer institute component of NSF's education program. "These are prime grounds for propagandizing our program," Zacharias noted, "and we must try to have materials ready for these institutes."[22] Given the finite number of high school physics teachers across the United States, Zacharias felt that a systematic communications effort could reach nearly all of them. The plan, he recalled, "was as deliberate as brain surgery." They started with 25 committed teachers, selected 12 "who wanted to teach the stuff as it came hot off the griddle," and then distributed those now experienced teachers among five summer institutes with 50 teachers each the following summer. In this pyramidlike scheme, as Zacharias explained it, "T1's teach students, and T2's teach T1's and students, and T3's teach T2's, T1's, and students." It was a highly efficient dissemination network that succeeded in reaching over half the physics teachers in the country by the late 1960s.[23]

Each component of the project was carefully coordinated, keeping in mind the various constraints and limitations that could impede reform at each step. At the Woods Hole conference, Fran Friedman described this work as the "*foreign policy* of education," that is, it was the question of "once you have got the ideas, principles, etc. How do you put these into effect?" He was specifically interested in how one goes about "managing the resistance . . . [of the] people in [the] school system who are responsible," which seemed at times a task as formidable as getting information behind the Iron Curtain.[24]

Top officials at NSF did what they could to influence the broader sphere of public opinion, to condition the public to embrace the education reforms being advocated by the scientists. The Cold War mindset of

Troy was evident from the beginning. At a meeting of the education division in 1956, one board member argued that the dominant ideology of public education, the "life-adjustment" concern with meeting all the personal and social needs of children, worked against the pursuit of more intellectual goals. The American people "are held captive by their own propaganda. . . . What is required here," he urged, "is counter-indoctrination, highlighted by people who know how to affect public opinion."[25] After it became evident that even the Soviet space successes had failed to create a sustained public interest in reforming science education, NSF director Alan Waterman sought to follow that advice. With the support of the White House, he turned to the National Ad Council, "skilled as it is in communication," to develop a public relations program that would remedy the public's "lack of realization and determination . . . to advance progress in education and in science."[26] In a lengthy memo, Waterman described the task before the council as one of using whatever media necessary "to translate this problem into terms which will provoke citizens to take quite specific action."[27] After a series of meetings of the Ad Council's Public Policy Committee, an "excellence campaign" designed to improve American attitudes toward intellectual pursuits was readied for implementation.[28] Along with the efforts of the President's Committee on Scientists and Engineers and the broader Goals for Americans initiative, this provided a key piece of the coordinated public relations agenda of NSF to manufacture as expeditiously as possible "a favorable climate in which to cultivate the learning process."[29]

In all these dissemination efforts, the scientists involved viewed the public primarily as an obstacle to overcome, as an impediment to the implementation of the new projects being developed—"resistance," in Friedman's words, that needed to be appropriately "managed."[30] Margaret Mead, in a roundtable discussion on the prospects for reform at the Yale conference, commented that she felt that, despite the existing hostility toward intellectual pursuits, the American public was "capable of learning and learning fast." At least as fast, she noted in all seriousness, as "some of the savages in the Pacific . . . have been able to learn."[31] This view of the public as an object to be manipulated could have been lifted directly from the final report of Troy. There, in the section dealing with the importance of mobilizing public opinion to support political warfare abroad, the authors explained that although "a major faith of the democratic state is that the people are inherently wise and just," this did not mean that the people should have "direct and literal control of public policy." The issues and strategies of such policy are just "too complex and remote for the people in the mass to grasp with justice and wisdom." What was important, the report argued, was that, within the limits of the fundamental attitudes and values people possess, public opinion could be shaped, and that research would reveal the various ways policy may be presented to ensure its eventual support.[32]

This was the path that PSSC, BSCS, and NSF chose to follow. Their models of dissemination were designed for expediency, not for reasoned deliberation of or public participation in determining the fate of education. To raise public levels of rationality in one sphere (the classroom), it was necessary—or so it seemed to them at the time—to subvert it in another (the general public). The techniques of mass persuasion that were so harshly criticized when linked to the life-adjustment curriculum in the early 1950s, found far more room to operate behind the scenes, out of the limelight of the classroom as part of the scientists' effort to smuggle in via their self-contained, neatly scripted curricular materials a view of science uncorrupted by faddish educational theory or the distortions of mass consumerism. Perhaps those involved felt that this was simply the price to be paid to gain a foothold for pure science in the classroom, where young minds could experience its inherent elegance, beauty, and power.

The techniques of social control and manipulation increasingly looked to in domestic policy and education circles were, without question, the products of the ideological battles waged during World War II and the ensuing Cold War. But this emphasis on technique that found such wide application in the social realm extended more deeply than scientists, perhaps, wished to acknowledge. The blurring of the line between ends and means that is evident in the project to disseminate sound reasoning skills, on close inspection, is evident as well in the very content of science itself—content that, in turn, became incorporated into the subject matter of the new school science.

## Physics at the Cutting Edge

The PSSC curriculum was viewed by scientists and educators alike as a radical break from traditional high school physics courses, not just in instructional approach, but in substance as well.[33] Zacharias's effort "to reformulate the subject in the framework of 1956 instead of 1896" produced a course that was nearly unrecognizable as physics to the typical high school teacher.[34] Gone was the physics of pulleys and levers, refrigeration, and internal combustion engines. In its place came waves, atomic energy levels, and the photoelectric effect—topics not previously found in high school physics classrooms, topics foreign to both teachers and students. But as strange as these subjects may have appeared to progressive educators of the 1950s, they made perfect sense to those looking out from the laboratories of the Cold War research establishment—the military-industrial complex described by Hersey. These modern concepts were the bread and butter of the physicist's working life in 1956; Zacharias was only attempting to bring the high schools up to date, to help students see just what the intellectual tools of that life looked like up close.

The most widely used high school physics books of the time, such as Holt's *Modern Physics*—which seemed to epitomize the "bad textbook" for the PSSC group, represented a very different educational tradition. As already noted, they were filled with gratuitous references to the era's modern conveniences, things such as refrigerators and automobiles, objects the physicists found to be distracting and even dangerous in the implication that physics and technology were one and the same, as many Americans had increasingly come to believe. But equally troubling to Zacharias and his colleagues was the underlying physics itself. The fundamental problem, in their view, was that school physics had simply failed to keep pace with research physics over the course of the previous 50 years. The passage of time since the early 1900s had resulted in two compatible, though radically divergent disciplinary frameworks. The predominant emphasis in textbooks such as *Modern Physics* was on what PSSC Steering Committee member Walter Michels referred to as "the understandable, complete and closed system of classical physics."[35] This consisted of the standard account of force and motion in Newtonian mechanics and the application of these ideas primarily to fluids (hydrostatics and hydrodynamics) and simple machines (e.g., pulleys, levers, and inclined planes), this in addition to extended units on heat and sound. All of these topics, Michels lamented, were more appropriate in an engineering curriculum than a physics curriculum.

As conceptually sound and intellectually revered as this kind of physics was, it remained, nonetheless, "quite at variance with the way in which *physicists* think."[36] At best, Zacharias noted, Newton's laws are "respectable *approximations* for bodies which are not too big, not too small, not too slow and not too fast."[37] Thus, at the cutting edge of research, where the MIT physicists lived and breathed, classical physics was an ineffective tool. The professional physicist "uses [it] only as a road that leads him to those matters which are of current interest," that is, "particles and waves," explained Stephen White.[38] This was, according to one of the early PSSC reports, "the physics which is being evolved by men of our generation; which is affecting [the student's] present and his future; which is still an open field whose many paths leading to the unknown he may elect to follow."[39]

This emphasis on particles and waves was the most distinctive feature in the content of the new physics curriculum. The two ideas made up a simple and elegant organizing framework for the entire course. The textbook opened with routine coverage of a number of introductory topics including time, measurement, and vectors, which provided students with the tools necessary for considering the conceptual material to follow. This began with an explanation of the atomic theory of matter and a description of the physical characteristics of atoms and molecules—things such as their ability to produce unique spectral patterns and the way they behave in the gaseous state. The PSSC writers then immersed the students in the subject of light and optics. The entire range of optical phenomena from reflection and refraction to the production of optical images was treated in depth—phenomena that were explained initially using a particle model of light. But after continued exploration, and with the help of various laboratory experiences, students were shown that this model failed to account for a number of observed behaviors. The wave model followed, which not only resolved the anomalous observations, but also provided the conceptual foundation for the rest of the course.[40]

Indeed, Zacharias and his group devoted a considerable portion of the instruction to wave phenomena. Alongside extended textual discussions of waves were numerous stop-action photographs of assorted wave forms followed by wave-related problem sets. To complement these items, Zacharias directed the PSSC apparatus group to design a cost-effective ripple tank—a device consisting of a horizontal pane of glass that supported a thin film of water into which a vibrating motor was placed in order to generate various wave patterns.[41] These patterns, which provided a ready analog to light wave patterns, were projected onto the floor or ceiling for students to analyze. In this way, as Zacharias explained it, the students were physically able to "put waves through their paces," a much richer experience than simply reading about and seeing them.[42] The goal of all the readings, laboratory exercises, and film demonstrations was not to teach students about the wave model of light per se, but rather to use light as a vehicle—"something within [the students] ordinary experience"—to convey the elements and power of the wave model for explaining phenomena related to electromagnetic radiation (of which light is one form) and matter.[43] This was perhaps the most foreign of all the ideas in the curriculum, but the one to which Zacharias was most committed. "All motions," he pointed out emphatically, "are guided by the laws of waves, not just light and sound. Everything that we know about moves as if it were a wave—you, me, the atoms, the electrons, everything."[44] The goal of PSSC, he explained later, was to get students to "have wave motion in their guts the way everybody has the motion of a ball in [their] guts."[45]

With this understanding of waves, students were steered directly toward modern physics—the physics of quantum theory and relativity, ideas of much more recent origin. Where traditional topics were included, they were subordinated to the dominant wave/particle theme of the modern point of view. Newtonian, or classical, mechanics, a staple of the prewar physics curriculum, was introduced much later in the course, and only then primarily to lay out the concepts of energy and momentum, which were abstracted from the familiar context of billiard balls and carts and applied to the invisible world of wave and particle phenomena. The conservation of energy and momentum, in particular, was central to understanding the properties of electromagnetic radiation. The material aspects of Newtonian physics—the carts that carried momentum, for example—were, in many instances, superfluous in this domain. They simply did not apply to the transmission of energy in wave form and were thus dropped from the picture.[46] Any effort to retain a Newtonian view, the physicists cautioned, would produce models that "are apt to be too complicated" and that "look artificial."[47] The course culminated in a detailed accounting of electromagnetic waves (which include the more familiar radio, light, and micro waves) and an exploration of the discrete interactions between these waves and matter (primarily absorption and emission phenomena) that is the heart of quantum physics.

Compared to the traditional textbooks, the PSSC course was not only more modern, but quite narrow in its treatment of physics, though the physicists would have preferred to call it unified and coherent. It would have been even more narrow had Zacharias followed through with his initial idea to build an entire high school course around a single molecular beam apparatus, an increasingly common piece of equipment that consisted of an evacuated glass tube roughly ten feet in length into which molecules or atoms were introduced at one end and detected electronically as they arrived at the other. By placing "a variety of slits or wires as obstacles, plates, magnets, or other things" in the path of the beam, Zacharias proposed to demonstrate the particulate nature of matter along with the properties associated with those particles.[48] Though the single apparatus idea was abandoned in favor of the coordinated instructional system approach, this fundamental conceptual core—particles (atoms, nuclei, and so forth) and their interactions with and behavior in electric and magnetic fields—remained the essence of the PSSC course. It focused on the atomic level because that was where physics was in the 1950s, which is why PSSC was initially conceived as a two-year interdisciplinary course that would span both physics and chemistry.[49] "The beginnings of atomic physics are very much to be found in chemistry," Zacharias explained.[50] The longstanding demarcation between classical physics and chemistry had little meaning in the atomic and subatomic world of postwar physical science.

Few would question that the physics of PSSC was up to date, that it was firmly grounded in the disciplinary paradigms of the time. The molecular beam had become one of the most productive tools of the practicing physicist; it was the preferred method for examining the interactions between particles and electromagnetic radiation. Zacharias, who enthused that "you can see molecular beams all over physics," was one of the leading authorities on their use, having spent most of the 1930s working with atomic hydrogen beams in I. I. Rabi's lab at Columbia University. [51] There he had been part of the team that had used radio frequencies (electromagnetic radiation with wavelengths in the meter range) to explore the fundamental properties of the hydrogen nucleus, work that earned Rabi the Nobel Prize in 1944. [52] Research along these lines was pursued throughout the physics community after the war. Right up to his immersion in PSSC, Zacharias had run his own molecular beams laboratory at MIT, where he developed techniques that resulted in the production of the first commercial atomic clock (a sealed, cesium-beam frequency standard) in the fall of 1956. [53]

This research focus on what were known as quantum electronic phenomena was not the result of some chance discovery, nor simply the next logical step in the conceptual evolution of physics over the 60 years since 1895. The events of World War II had brought this particular branch of physics to the forefront of the scientific research establishment. The concerted efforts of physicists to expand their understanding of and exploit quantum effects was the direct result of their success developing the new military technologies that utilized microwave radiation; the most prominent examples were, of course, those developed during the war—radar, the proximity fuze, and so on. With the fission techniques worked out at Los Alamos thrown into the mix, the war had opened up bountiful new research fronts, which had significantly reshaped the research practices of physics, not just in the modes of organization or the analytical systems approach to attacking complex problems as we have already seen, but even in the types of questions the physicists chose to ask. The war had given them new material apparatus, a unique blend of pure research and engineering skills, and the funds necessary to exploit the esoteric world of quantum phenomena. At war's end, the physicists were, of course, free to pursue whatever sort of research they wished. Their wartime experiences, however, had primed them to continue down the path they had already begun. [54]

Without question, the most significant force driving this transformation was the unprecedented levels of government funding. From just before the war in 1938 to 1953, overall funds for physical science research increased over twentyfold. And, despite the repeated calls for greater funding of basic science, in 1951, according to historian Paul Forman, 70 percent of all academic physicist and senior research personnel time was spent on defense-related work. More startling is the fact that in 1954, 98 percent of the federal research money provided to academic physicists came through the Department of Defense and the Atomic Energy Commission, the institutional heir to the Manhattan Project, the primary business of which was nuclear weapons development.[55] With government involvement of this sort climbing ever more rapidly as the space race got underway, even the public seemed to notice the changes taking place. An editorial in the *Nation* in February 1958 noted the symbolic union between the scientists and their patrons in the form of a White House "science-military dinner." "Already the public," the magazine noted, "is acquiring a new image of the scientist—not an egghead but the man who makes missiles for the military, that is, a military man minus a uniform."[56]

In the face of this overwhelming military patronage, scientists tried to maintain clear distinctions between their basic scientific work and the contract research they conducted on behalf of their military sponsors.[57] The military funding agencies even provided a good deal of nonprogrammatic money that physicists used to pursue basic research-type projects.[58] Thus, the illusion of intellectual autonomy—the freedom from government direction—was largely maintained.[59] But external appearances and personal perceptions notwithstanding, these various governmental agencies were clear from the beginning about the returns they sought on their investments. James Killian himself, who as MIT's president was well aware of the ins and outs of military subsidies of academic science, explained that although "the Department of Defense . . . has steadily supported the concept of basic research," such support only came "in those areas where the advance of knowledge is important to the military effort." In the early 1950s, he noted, the DOD had affirmed that it was indeed their policy to provide for "basic research performed as an integral part of programmed research committed to specific military aims and for . . . academic research that promises ultimate military application."[60] Most of MIT's famed postwar labs such as the Research Laboratory in Electronics (RLE) (the postwar version of the Rad Lab), the Laboratory for Nuclear Science and Engineering (LNSE), and the Lincoln Lab were built on this funding, the research conducted there contributed to these military goals. Zacharias's professional life was deeply embedded in this institutional matrix both as a researcher at the RLE, where he developed his atomic clock, and as director of the LNSE.[61]

In ways sometimes subtle and sometimes not, this flood of money inevitably pushed academic physicists as a whole to pursue research that ultimately served the interests of the national security apparatus. The result was scientific work, both basic and applied, geared toward the enhancement of man's ability to manipulate and control natural processes. In the words of Paul Forman, who has argued this point convincingly in his path-breaking historical work on the federal patronage of postwar physics, "support by military agencies and consultation on military problems had effectively rotated the orientation of academic physics toward techniques and applications."[62] The field of quantum electronics, he notes, was exemplary in this regard. Within this field, physicists used defense department money to explore the nature of the various interactions between electromagnetic radiation and quantum systems. Subfields such as microwave spectroscopy, electron spin resonance spectroscopy, and nuclear magnetic resonance spectroscopy burst onto the scene in the decade following World War II.[63] Though conceived as areas of fundamental research, the pursuit of which earned countless students their Ph.D.s and senior physicists Nobel prizes, all of these areas of inquiry focused more on extending the range of human mastery over nature than on the development of theoretical understanding.[64] One physicist surveying the field of microwave research in 1958 admitted as much. "Whereas, . . . one can study cosmic particles in the hope that this study will shed new light on the basic properties and laws of nature," he commented, "the study of microwaves in the same sense is pointless." "This does not imply," he went on, "that research which makes *use* of microwave techniques in not possible or profitable— quite to the contrary." It was simply important to recognize, he explained, that research in this field "must, of necessity, be *applied research,* always directed toward a useful goal in some other field."[65]

The emphasis on quantum electronics was naturally fine with the defense agencies providing the funds. The techniques and conceptual knowledge found wide application in key military technologies—technologies used for more sophisticated systems of ship and plane navigation, detection of enemy craft, and both field and intercontinental communications. Even Zacharias's seemingly innocuous atomic clock soon became a key component in advanced nuclear missile tracking and guidance. (That it was funded almost completely by the military should come as no surprise.)[66] Many of these new applications were conceptualized in the various defense summer studies that Zacharias had pioneered. The participants of Project Troy, for example, devoted a notable portion of their study on counterpropaganda to the problems of radio transmission and jamming over the European continent. Here was a theater of conflict almost completely reserved to the scientists and engineers who were experts in the physics of radio wave propagation. There was the potential in the jamming and counter-jamming of such transmissions for what the Troy participants called an extended "electromagnetic war" with the Soviets.[67] The Distant Early Warning Line (DEW), a string of radar stations in northern Canada deployed to warn of a Soviet missile attack, was yet another example of how the new physics could be fruitfully applied to military concerns.[68]

Ironically, the rotation of academic physics toward techniques and applications, specifically those related to quantum phenomena during the Cold War, had the opposite effect on the substance of high school physics. In order to bring the subject up to date, Zacharias and the rest of the PSSC physicists, in addition to eliminating the physics of everyday life, passed quickly over classical physics so that they could fully treat the modern topics researchers tackled at the behest of the military. The result was the further displacement of the real objects of the world—if only as real as the wooden force carts and billiard balls of traditional physics—with the more intellectually demanding and highly theoretical entities that were subatomic particles and electromagnetic fields. Though foreign to many high school teachers and certainly to most students, these new objects of study that constituted the particle/wave theme of the PSSC course were warmly familiar to practicing physicists of the time, Zacharias's friends, colleagues, and fellow summer-study participants. After all, PSSC was a perfect high school primer for the new decidedly instrumentalist physics of their time.

## Biology in Transition

Like the new physics, a new biology had emerged within the United States during the middle decades of the twentieth century. The conditions of its development were more complex than the comparatively straightforward shift toward technique experienced by the physicists. Though similarly influenced by the changing political economy of postwar government research policy, the biological sciences, as we have seen, also agonized over their disciplinary status in the years following the "physicists'" war. As a result, biological research, despite the best efforts of biologists to resist being subsumed by physics and chemistry, increasingly patterned itself after research in the physical sciences. Biologists became favorably disposed to using complex instrumentation, participating in large-scale research projects, and, most significant, pursuing theories and models that in their very construction facilitated the human manipulation and control of living systems.[69] But unlike the situation in physics, where curricular updates were widely accepted as long overdue corrections to an outdated school subject, biologists struggled with the question of just what an accurate, up-to-date high school biology course should look like. At one level, the production of the three textbook versions reflected the diversity of viewpoints that seemed to plague biology through-out the century. Yet, at a more fundamental level all of the BSCS curricular materials ultimately were united, however loosely, by the technique-centered research ideology adopted so successfully in physics.

Though it is difficult to cleanly separate the core biological content from the various social agendas and pedagogical accommodations in the pre-BSCS textbooks (particularly at the high school level), most practicing biologists interested in education were in fair agreement that, as far as content went, introductory courses devoted too much time to the main taxonomic categories and the life forms they contained.[70] "In the old biology course," as John Moore characterized it, "you went over all the phyla of plants, and then all the phyla of animals; at the end, if you had time, you had perhaps two lectures on Mendelism, one on embryology, and maybe one on ecology and importance to man."[71]

This aptly described the material in most high school biology text-books of the time. The authors of *Modern Biology,* the most popular of these, opened with a brief section on the nature of biology and life, delved into classification systems, and then promptly began the rather dry march through the various categories of living organisms, beginning with plants and moving up the scale of complexity from single-celled organisms all the way to animals with backbones—the most significant, of course, being "man." These sections were, in the spirit of the life-adjustment movement, interspersed with prescriptive material on human hygiene, behavior and personality, and the assorted applications of this knowledge in everyday life.[72] The biological content itself, though, remained primarily descriptive, in the tradition of classic natural history. Providing this comprehensive sur-vey of the living world, many believed, would convey to students the grandeur of life. It would allow them to appreciate, in the words of Illinois botanist Harry Fuller, "the basic, the impressive, the truly significant bio-logical phenomena: the panorama of life through geological time, the mar-velous interrelation of tissue structure and function in living bodies, the wondrous adaptations of flowers to pollinating insects, the mysterious mi-grations of birds," and so on.[73]

When the biologists sat down to overhaul the high school curriculum, everyone agreed that modern biology was much more than the grand cat-alog of life commonly portrayed in introductory textbooks. Missing were the dynamic, experimental aspects of modern-day science. The physical scientists had demonstrated the fruitfulness of the new type of research—research that was fully embraced and promoted even as it was, in turn, shaped by the state. For biologists, the path to status, funding, and influence seemed obvious, and it was one that many had chosen to follow. It entailed importing quantitative methods and experimental techniques from the physical sciences. With these tools, biologists along with a growing number of physicists, disillusioned with the recent horrors mastery of inanimate nature had produced, established a variety of new fields of study at the border between biology and physics. The new interdisciplinary fields included ecosystem studies, regulatory biology, population genetics, and, at the forefront of them all, molecular biology, which began its disciplinary life as "biophysics."[74]

In the new biology, field observations and anatomical description were pushed aside to make room for the cutting-edge conceptual and material tools of calculus, systems analysis, electron microscopy, radioactive tracers, viral and protein crystallization, cell irradiation, and x-ray crystallography. These tools brought with them the social authority of physics and inevitably, as historian Evelyn Fox Keller has argued, in their deployment on living systems recast the very goals of biological science from observation and description to intervention and control.[75] By the late 1950s, biology was becoming, as physics had before it, a science of technique rather than one of understanding.

Though reconstructed internally, this new hard-research biology was still not widely noted or appreciated among those outside the field. A well-designed education program, BSCS members believed, would go a long way toward changing the prevailing view among the public and funding agencies that biology was merely "a descriptive science but not a basic science." Steering Committee member Wallace Fenn insisted that biologists must dispel the notion that "the real fundamentals of biology are the responsibility of the theologians and philosophers and have no place in the realm of the experimental sciences." In a speech at the annual meeting of the American Institute of Biological Sciences in 1960 almost childishly titled "Front Seats for Biologists," Fenn made his case. Fighting to place the discipline on par with physics, he explained that biology "was experiencing a revolution which may turn out to be as dramatic as the revolution in physics. The advances in nucleic acid chemistry have greatly illuminated the structure of genes and chromosomes to such a degree that growth and development and inheritance . . . become almost understandable. Viruses can almost be synthesized in the laboratory. Several reasonable explanations for the origin of life have been offered." And, drawing explicitly on the epistemological authority of physical science, he asserted that "the modern biology department is not complete without someone who knows biochemistry and biophysics and mathematics and is exploring fundamental problems in cell function. Modern biology is a dynamic analytical science and not merely descriptive."[76]

This was the bias that the BSCS group brought to the project in its initial phase. At a meeting of the film committee, which worked closely with the curriculum group, it was stated that "It is . . . the intention to create a strong course which places more emphasis on contemporary biology than has been common at the secondary school level in the past." The "matters of emphasis" identified by the committee included: general, cellular, and comparative physiology; interrelationships, energy and matter circulation patterns; adaptation as a dynamic process; and scientific methodology rather than static morphology and taxonomy.[77] All of these topics fell right in line with the new paradigm of biological research that emerged in the wake of the physicists' success. They were an attempt to capitalize on the fact that, as Moore declared in a memo on the objectives of the curriculum project, "the accuracy of prediction and control of biological events . . . is great and becoming greater."[78]

When it came time to put the new books and curriculum materials together, though, the strong hold of tradition made a clean break difficult to come by. Early on, some of the biologists suggested that modern experimental biology might be too difficult for high school students, arguing that at the tenth-grade level the content should be primarily descriptive.[79] A plea for retaining the descriptive aspects of biology as a matter of principle rather than pedagogy was made by the American naturalist and writer Joseph Wood Krutch, who, having learned of the BSCS undertaking, wrote to Grobman to make a case for a humanistic biology. "I feel very strongly," he stated, "that there is still a place in a liberal education for the sort of familiarity with the natural world which does not come from the laboratory, from dissection or experiment." Krutch, who later joined the Steering Committee, insisted that such an education is "incomplete if it does not include some familiarity with the living world outside man as a subject of wonder, delight, and aesthetic satisfaction."[80] But every plea for the descriptive seemed to be matched by equally strong pleas for the experimental. Not long after Krutch's letter, Steering Committee member C. Ladd Prosser, a physiologist from Indiana, expressed his concern that the curriculum "place considerable emphasis on dynamic biology and perhaps a little less on descriptive biology than in existing biology courses." "I do not imply that we should in any sense omit descriptive biology," he explained. "what I really mean is that I see no need to wait . . . to present the dynamic aspects of biology and the physico-chemical approach to life."[81] Such conflicting sentiments reflected what one biologist called the growing "schism between molecular biologists, biochemists, etc. and old style biologists."[82]

There certainly was no indecision at NSF about where they felt biology should be headed. As Toby Appel has shown in her history of the Foundation's biology and medical science division, key program officials pushed for the adoption of the more experimental, molecular approach in all the fields of biological study.[83] There was the sense at NSF that the newer techniques brought over from the physical sciences were just the means by which the debilitating fragmentation of biology might be overcome. In sketching out the future of biological research, one Foundation official wrote, "the concepts of genetics, biochemistry and biophysics are unifying influences and we are presently witnessing the transformation of the life sciences from a discipline-orientation to a unified science."[84] The basic tools of the biology laboratory were no longer "a light microscope, dissecting tools, pins, rubber bands, wax, clay, etc." The new era of research, as NSF biologists saw it, required "ultracentrifuges, amino acid analyzers, electron-microscopes, recording spectrophotometers, electrophoresis and chromatography apparatus," and the like.[85] The goal of the directors of the Biological and Medical Science Division was to structure their patronage in ways that would "speed the applications of new ideas and new techniques, to sharpen the powers of resolution" in all the disciplines of biology.[86]

The progress of BSCS toward this vision, under the guidance of the Foundation's Division of Scientific Personnel and Education, appeared to outsiders to be halting at best. So much so that the curriculum group was used by biology division officials at NSF more than once as an example of the complete disarray in which biology found itself at midcentury. In one instance, the assistant director for the Biological and Medical Sciences Division in reporting to Alan Waterman on the state of life-science research in the United States wrote: "Biology faces important challenges, for the discipline as a whole is at a crossroads. There is so much disagreement today as to what biology is that a group recently trying to design a new high school course, like the blind men describing the elephant, came out with three distinct versions."[87] Arnold Grobman was understandably upset by this characterization of the project, particularly coming from within the agency (though not the division) responsible for its funding. "I deplore the fact that a biologist serving as a high ranking official in the NSF," he wrote in a memo to the Steering Committee, "apparently does not understand why we have produced three versions." This was all the more astonishing to Grobman since BSCS was a project "in which the NSF [had] invested six million dollars" by the end of 1962.[88]

Grobman's views notwithstanding, the program officials in the biology division can be excused for seeing the BSCS project as conflicted. On the surface, the three versions did appear to be a poor compromise designed to appease competing disciplinary factions. Though the blue version reflected the dynamic, molecular approach that the Foundation was looking for, the green version appeared to offer students and teachers the descriptive, natural history treatment of biology to which they were more accustomed. It was a version that represented, according to those in the Foundation's biology division, though, "the lack of vision . . . in many of our universities" for cutting-edge research.[89] The yellow version looked to be little more than a variant of existing high school biology textbooks with some added emphasis on cellular processes along with a dash of history. Clearly, the internal deliberations of the Steering Committee tend to support such a reading of intellectual pluralism, if not disarray. However, something more than a cursory examination of the materials produced would have left those in NSF's biology division more comforted than critical.

Certainly no one could take issue with the blue version, which fully embraced the "biochemical-physiological slant" advocated by NSF.[90] But in nearly every other component of the program, one could also see the shift away from the old-school biology to the new reductionistic, experimentalist mode of biological research. The initial list of supplementary films to be used with all three versions of the text, though including a handful on the various biomes of the world, was weighted heavily toward experimental biology. They treated topics rarely found in high school classrooms such as hormonal control; the isolation, purification, and identification of organic compounds; microdissection techniques; the effects of radiation on chromosomes; and nuclear transplantation in frog eggs, to name only a few.[91] The intensive lab blocks developed for the course similarly favored the experimental over the descriptive. Though Bentley Glass claimed that the idea of the laboratory should encompass the field as well as the lab bench, 9 of the first 13 blocks emphasized controlled laboratory experimentation with only one devoted to field study. Among the very first of these were labs on the regulation and control of growth and development in plants and animals, highlighting in particular hormonal manipulation of test organisms.[92] The teacher's handbook produced by BSCS, in addition to containing Schwab's invitations to inquiry, devoted over a third of the book to nine detailed treatments of various topics in physics, chemistry, biochemistry, and statistics in an effort to provide the necessary hard-science background biology teachers would need to teach the new course.[93]

Even the areas and topics BSCS writers brought in to represent traditional biology bore the unmistakable imprint of the new Cold War research culture. The more descriptive type of work that was common in fields such as taxonomy and classification and ecology had undergone subtle transformations after the war. Biologists who studied taxonomic relationships between organisms, for example, began to add the tools of biochemical and molecular analysis to their more antiquated inventories, checklists, and detailed morphological descriptions.[94] There is no evidence to suggest that the emphasis on the newer techniques and approaches as they applied to traditional, descriptive biology in the BSCS curriculum was the result of a calculated effort to recast classical biology in the postwar research idiom. Rather, as was the case with the physicists, it seems to have been more the result of the Steering Committee's desire to fashion the new high school course in the image of current research practice. The new techniques and outlook were incorporated simply as a matter of course.

Perhaps the most striking example of this sort of reorientation can be seen in the green version. Of the three books produced, it was widely seen as taking the descriptive, natural history approach to biology that Joseph Wood Krutch had so strongly supported.[95] In the book's opening pages, Marston Bates, the primary author, introduced students to the study of living things with a rabbit sitting under a raspberry bush rather than with atoms, electrons, or the more complex chemical building blocks of life. Bates was well known for his writing in natural history, publishing a number of books on topics from the natural history of mosquitoes to the nature of natural history itself.[96] But even he had recognized that changes were afoot in the traditional domains of biology. In contemplating the state of natural history, he noted that we "have been so busy with the enormous job of observing and describing life phenomena that we have had relatively little time to spare for experimenting with the modification of the observed processes." The future, he believed, lay in moving toward greater intervention. "The finding of appropriate experimental methods," Bates explained, is "the most important task confronting natural history." After all, it seemed increasingly obvious to him that "the progress of science is all bound up with the progress of instrument manufacture and of manipulative technique."[97]

The green version was subtitled "an ecological approach," and though Bates had insisted that natural history and ecology were really one and the same, by 1959 ecological studies in the United States were operating with an entirely different set of conceptual and technological tools from those used by natural historians early in the century. After World War II, ecologists, who were generously funded by the Atomic Energy Commission (AEC), used gamma rays and other forms of radiation to massively disrupt ecological communities in order to study how they functioned and the ways in which they might or might not recover from such devastation. Paleoecologists used radiocarbon dating in their analyses of pollen in pond sediments. And the use of radioactive tracers proliferated in studies of nutrient cycling in complex ecological communities.[98] Perhaps the most profound change, though, was the gradual adoption of the view of biological communities as complex, self-regulating feedback systems. The systems metaphor, as we have seen before, held immense appeal in the years after the war primarily for the opportunities it offered for the rational management and control of nature. Central to the work of ecosystems studies was the analysis of the flow of energy and matter through the system, a system that encompassed not just biological factors, but now chemical and physical factors as well.[99] Energy flow diagrams and nutrient cycles filled the opening chapters of Bates's "natural history" textbook. Though students began with the pastoral image of the rabbit under the raspberry bush, a few pages later they would find the rabbit and bush assigned to their appropriate trophic levels performing their functional role in an elaborate ecological feedback system.[100]

Evolution was, of course, the theoretical framework that the biologists believed would unify the various components of the course as well as the fragmented practices of biology itself. It too exhibited the characteristics of the new approach. While in the past, evolutionary arguments were grounded on the static evidence derived from lifetimes of collecting and cataloging descriptive data common in fields such as comparative anatomy, biogeography, and paleontology, developments in quantitative population genetics and methods of molecular analysis gave evolutionary biology new life as an experimental science.[101] In this molecular genetic form, evolutionary biology not only enabled, but even invited external direction and control. It was a view that, in Hermann Muller's words, saw life as nothing more than a "peculiar chemical set-up which results in replication, mutation (non-adaptive) and, by natural selection of the mutations, the evolution of the diverse and highly complicated adaptive systems"—a view he believed high school students needed to appreciate.[102]

For both Muller and NSF, the future of biology lay in the ability to manipulate that chemical/biological process. The frontiers of such research included: "the molecular basis of thought, memory and learning," along with "the problem of aging, . . . [the] controlled modification of the genotype (genetic engineering) and control and regulation of the phenotype (euphonic engineering)."[103] However dissatisfied some research biologists might have been with the BSCS project, the new high school biology material it produced contributed far more than it detracted from the discipline's effort to sell itself as an important player in the government-science complex, as a discipline capable of generating instrumental (rather than descriptive) knowledge that could sustain biologists' claims of relevance in the new technocratic order.

What makes these transformations all the more remarkable is the fact that they were occurring during a period of heightened professional consciousness on the part of scientists. BSCS biologists, in particular, were keenly aware of the potential for disciplinary distortion from external factors. This was evident in a disagreement between Glass and Grobman over the desirability of seeking funds from the AEC for the development of classroom materials related to the topic of radiation biology. Glass, who had initiated contact with the agency, felt that teachers and students would benefit tremendously to have materials "prepared under the auspices of the AEC and with its financial support."[104] Grobman, however, sensitive to the influence of outside interests, objected: "I do not think a project designing educational materials for use in the public schools should be supported in any way by quasi-military or military agencies." He was confident that the support they had been receiving from NSF, despite their past heavy handedness, could be considered "clean." "By that I mean that we have had absolutely no interference regarding the materials we have designed. We can truly say that our materials represent the thinking of biologists and are not influenced by the goals or special interests of any of our supporting foundations."[105] Yet, despite Grobman's faith in the purity of their work, biology was being remade by such interests nonetheless.

It should be emphasized that neither PSSC nor BSCS were in any significant way responsible for the transformations in scientific organization and research practice that became firmly established during the Cold War. The curricular materials the scientists had produced simply reflected what the scientific enterprise looked like in the late 1950s and early 1960s. What is interesting is the incongruity between the aspirations of the reformers and the results of their work. Contrary to the perceptions of the time crystallized by Hersey in *The Child Buyer,* as well those of historians today, the hothouse production of a new generation of scientists to do battle with the Soviet Union played little if any part in the scientists' grand social plan. Their self-professed goal was the development of public understanding of science in its purest sense, without the confusion that the gadgets and technologies would bring. They wished to cultivate a social milieu in which the intellectual pursuits of scientists were both respected and supported, even deferred to.

In looking at the reforms produced, one might say that they succeeded in that effort, but only partly so. Students using the new curriculum materials would gain an appreciation of what research science was all about to be sure. But, with the changes in the character of knowledge wrought by the Cold War, it would be less about science as a creative, humanistic pursuit as the scientists intended, than about the interventionist research ideology that had come to permeate nearly all scientific disciplines. As much as scientists wished to turn public thought "away from gadgets, rockets, antibiotics, plastics," and the like, as Warren Weaver had insisted should be done, the new form academic science had assumed with the help of its government patrons made it all the more possible to bend nature to the needs of the state. American science in the years after World War II had in many ways become one with engineering.[106] This was the "pure" science students would see in the classroom experiences of PSSC and BSCS. In this sense, the scientists had accomplished their goal: the students of the 1960s were without question brought up to date.

CONCLUSION

# POLITICS AND
# THE SCIENTIFIC IMAGE

In the opening pages of the *Biology Teachers' Handbook,* Joseph Schwab looked back over the history of science textbook publishing in the United States and described three phases. The first was from 1890 to 1929, during which time the content of school subjects reflected the structure of their corresponding academic disciplines. Though these textbooks consisted mostly of "disconnected facts and elementary generalizations," they were, nonetheless, written by "working scientists or colleagues of working scientists." The second phase, from 1930 to 1960, saw a rapid increase in the size and diversity of the student population and a corresponding rise in the influence of the professional educator on the production of science curricula. The content of the textbooks of this period, Schwab noted, "was no longer mainly determined by the state of knowledge in the scientific field. Important ideas "were omitted or emphasized on the basis of views as to what could be most readily taught," and what remained was often "modified to conform to theories of teaching and learning," he explained, "regardless of the extent to which these modifications presented a distorted view of the subject as known by the scientists." The third phase, of course, had just begun, characterized by the awakening of scientists to their responsibilities for educating the public to the nature of science in the new age. The cycle was thus complete.[1]

What is most interesting for the purposes of this book has been the shift from phase two to phase three, a short span during which scientists reasserted their authority over the content of the pre-college curriculum. Questions regarding the nature of this transition have gone unexamined for too long. Why, for example, were the scientists so quiescent for the 30 years that the educators shaped the nature of classroom instruction? And what prompted them in the mid-1950s to concern themselves suddenly with high school science education? Just why did, in Schwab's words, "large numbers of scientists, after years of indifference or disdain for educational problems, . . . come out of their laboratories, in person and through their societies" to call for the reform of science teaching?[2]

As I have attempted to demonstrate in the preceding chapters, the source of the scientists' interest in education reform, contrary to the conventional accounts of these events, had less to do with either the activities of professional educators or the technological threat posed by the Soviet Union than has been supposed.[3] The reasons the scientists decided to volunteer their services to the cause of education reform were fundamentally political and tied directly to the rapid integration of science into the national security infrastructure of the United States. During the 30-year reign of the educators, if we accept Schwab's periodization, science as an institution devoted to the discovery of the most esoteric knowledge of the physical world had little interaction with the federal government, at least outside of the field of agriculture.[4] Cloistered in their laboratories and in the employ of largely self-sufficient colleges and universities, scientists simply had no reason to have more than a passing concern with the public perception of what they did. The war, of course, changed all that. Scientists enlisted in the national security effort, and the federal government invested heavily in the scientists. The result was a tightly integrated institutional arrangement of enormous scope that had previously been nonexistent.[5]

It was in this context that they "awakened" to their interest in the schools. The new political environment in which scientists found themselves brought unforeseen pressures to their work: a push for applied research, preferential funding for select scientific disciplines, greater consumer expectations, and more government interference than ever before—restrictions on communication and travel, loyalty oaths, and reckless anticommunist investigations.[6] Scientists began to appreciate that "the climate of opinion about intellectual activities influences the progress of science, the uses of science, and the kind of people who enter science."[7] It was therefore not a coincidence that those who had traveled so extensively within the inner circles of the government-science nexus—Waterman, Kelly, Zacharias, Friedman, Killian, Glass—were the same individuals who initiated the radical reform of science education. Membership on curriculum steering committees multiply overlapped with the various governmental advisory committees and the leadership rosters of the national scientific societies. The acronyms—PSSC, NSF, PSAC, BSCS, NAS—all began to run together.

As we have seen in the planning and development of the new high school science courses, the objective of these scientists-turned-educators was to convey a more accurate picture of their work to a misinformed public. But what "more accurate" meant was conditioned by the Cold War environment and the interests of the scientific community operating within that environment. As the steering committees worked to define their disciplines, they struggled with the constant tension between science and technology. The application of science during the war was what won them the influence they enjoyed. And it was the Cold War-inspired fear of Soviet supremacy in scientific manpower that provided their opening into the schools. But as important as this public appreciation of the scientists' technological prowess was, it was an appreciation they sought continually to redirect toward what they saw as their real work: basic research. The nature of this redirection, of course, varied. The physicists had the luxury of divorcing their presentation of physics completely from its applications, while the biologists felt the need to stress the importance of basic research in their discipline for understanding the new wave of social problems crashing to the fore in the early 1960s.

The tensions between pure science and technology in the course materials paralleled a similar tension, though rarely discussed, between the technocratic means the scientists employed and the liberal scientific worldview they hoped to disseminate. It is ironic that the scientists, while espousing the values of intellectual freedom and creativity, the importance of reason and deliberation in society, worked, in many ways, to routinize and restrict the intellectual role of the teacher through the careful construction of their curricular systems, even to the extent of replacing the teacher completely in some instances, as in the film program of PSSC. This technocratic perspective extended to NSF as well as agency officials attempted to bypass much of the deliberative process of local determination of school content in response to the seemingly intransigent bureaucratic system of teacher education and the decentralized organization of schooling. Foundation officials strategically targeted practicing teachers and the school curriculum as the "weak" points where they could exert the greatest influence.

One might argue, as the scientists undoubtedly believed, that the means justified the ends. But as grand as their educational goal of developing the "rational man" might sound, it was here that the scientists took a step back from their professed commitment to universal reason. Though the new courses represented an improvement of sorts over the life-adjustment education program, something akin to Bestor's vision of disciplined intelligence for all, they also entailed a fundamental curtailment of the Deweyan ideal of expanding scientific thinking to all facets of human affairs.

Science certainly was not presented as a method to be used in solving the problems of daily living, though it was hoped that students would learn to recognize sound reasoning when they saw it (or, more importantly, reckless and unsubstantiated claims, such as those thrown about during the McCarthyite witch hunts that so disturbed Zacharias). The primary reason the physicists and biologists opened up their laboratories and disciplines to the public through these textbooks, films, and labs was to show the public what science "really" looked like. They sought to display the kind of knowledge that pure research generated in order to reduce the unrealistic expectations for the magic or miraculous. They offered students a taste of the intellectual process professional scientists engaged in to demonstrate the legitimacy of scientific methods for arriving at reliable knowledge about the natural world. One might say that the public was brought into the laboratory in order to keep them out in a preemptive move of sorts. For what the public really saw was not the messy, conflicted workplace populated by researchers, administrators, and military contractors, but rather a tidy little anteroom arranged to look like a laboratory housing scientific work as the scientists imagined it could be, or should be. The point was to make sure that visitors would not be inclined to wonder what lay behind the far door.

What the scientists of PSSC and BSCS ultimately sought to build was institutional trust, a faith in the expertise that professional scientists possessed.[8] By providing a window into the anteroom of the scientific community—about which were displayed the noble goals of truth seeking, reliance on evidentiary reasoning, and commitment to democratic self-regulation—the scientists intended to provide a reasonable basis for the public to continue providing financial support for research and, at the same time, to free them from oppressive government interference and control. Such a science-friendly environment, they felt, was crucial to U.S. survival in the tense bipolar political world that accompanied the new scientific age. Philip Morrison understood this imperative without question. He recognized the importance of acknowledging the institutional needs of the postwar scientific complex, that science was now a permanent "part of society." "Here is perhaps the greatest gap in our present scientific curricula," he warned. "The self-assured justification of science, that knowledge is always good in the long run, remains true; but the number of those who are skeptical is growing. It is up to the scientist to show [the public] that his faith is sound."[9]

Any hope to develop this trust required a fundamental shift in the nature of science education. The new focus clearly had to be on the scientific community itself and on science as a highly circumscribed disciplinary practice with legitimate (and powerful) methods, but only as they applied to a relatively narrow domain of natural phenomena, and only by those few properly apprenticed in the research craft. The previous educationist emphasis on science as a nebulous, disembodied attitude or method that would provide meaning and personal direction for each and every member of society diverted attention away from scientists and opened science up to political interference. It was in this sense—the limiting of both the range of application and those who might be capable practitioners—that the reformers took a step backward from the "science for all" ideal. Of course, rather than seeing this constriction of scientific practice as an abandonment of the democratic ideal, the scientists would have described it more as a reappraisal and redirection of the intellectual resources of science at a time of national crisis—science for the few in the *best interests* of the many.

The process of image making is inevitably revealing, and it is particularly so in this instance. The scientists' educational efforts to refashion the public image of science in the late 1950s is telling along a number of lines. Their discussions and actions reveal the concerns and anxieties they felt professionally. They highlighted their faith in rationality as a basis for human progress, their commitment to education as an effective agent of social change, and, of course, their view of what science was at its very core. The historical account detailed in the previous chapters provides a sound illustration of how fruitful the field of science education can be to historians and sociologists of science seeking insight into questions about the place of science in society, or to those scholars interested in more specific questions related to public understanding of science. The institutions and practices associated with formal education at the pre-college level have only just begun to be cultivated as an area of study.

There are important insights to be gained for educators here as well, both policy makers and practitioners. More than anything, the story of these NSF reform projects illustrates the highly contextual nature of schooling. It provides a concrete example of the just how pervasive social and political factors can be in shaping the educational infrastructure of this country. The World War II techniques and weapon-system metaphors, as we have seen, were central to the scientists reform efforts. They provided a model for reform that not only defined the solution to the educational problems of America at midcentury, but also what counted as a problem to begin with and the means to be employed in its solution. Of course, at one level, the susceptibility of education to external forces is nothing new. The influence of industrial models of scientific management early in the twentieth century, which produced among other things the platoon system in Gary, Indiana and activity analysis as a method of curriculum development, has been well documented in the history of education literature.[10]

The fact, then, that wartime R&D models and systems-engineering techniques were transferred from military activities to the problems of education, though overlooked until now, is not all that surprising, though much might be said about how these particular metaphors and models have persisted and influenced our conception of education down to the present.[11]

What has been neglected in historical studies of science and education, though, is an appreciation of the extent to which the disciplinary subject matter itself (over and above the specific process of curriculum development or the organization of the school day) has been shaped by external forces. It seems as common to speak of science today in the same manner as Nicholas DeWitt did in his book on Soviet manpower in 1955, or as Eisenhower and CIA director Allen Dulles did during their various NSC meetings, as an activity that possesses a remarkable intellectual integrity, that is somehow immune to political and social influence or direction. It was this very perception of ideological purity, as I have tried to demonstrate, that in part gave the scientists and the federal government their entrée into the field of educational policy and allowed them to remake the high school science curriculum as they desired. But even as they traded on this perception of science as free from corruption, even as they sought to bolster it further through their own public relations initiatives and education reform efforts, the ideal somehow slipped through their grasp. Science, though never really immune in the way scientists or the public thought, by the early 1960s was no longer what it once was, having been reshaped by the military-industrial interests that had become equal partners in the pursuit of knowledge and technique.

In some sense the cultural advancement project to which the scientists directed their new courses was turned on its head. As much as Zacharias, Friedman, Grobman, and Glass sought to remake school science into a force that could push or nudge mainstream American culture toward their own worldview, one that embraced rationality and appreciated the expert role scientists would need to play in the future, American Cold War culture ultimately remade them and the essence of the research that they did. As they worked to illustrate for students what scientific research was in the curriculum materials they produced, at the same time, in the very same materials, they illustrated how much science had changed, how science education had become unwittingly an instrument for American Cold War foreign policy. Science, in this case, was materially transformed by the social and political forces of the time—the science curriculum, as a stylized representation of science, even more so.

Of course, these various influences are never present in isolation. What emerged in the curriculum materials was an amalgam of interests and ideas.

The products, as a whole, cannot be understood apart from the historical context in which they were generated. The PSSC content emphasis on waves and molecular beams was the result of the disciplinary shift toward militarily useful techniques, as was illustrated in the last chapter. But this was joined by other interests as well. The physicists' deployment of scientists on film, for example, reflected their judgment of teacher incompetence and the need to craft the image of the scientists as nonthreatening, accessible, and unquestionably American in the midst of widespread popular suspicion of communist subversion. The inclusion of evolutionary biology in the BSCS curriculum can be understood as both a tactic in the postwar disciplinary competition with physics and a jab at the ideological control of genetics and evolution in the Soviet Union and religious conservatism in the United States. And the numerous activities and intellectual space devoted to scientific inquiry emerged from the scientists desire to showcase the intellectual power of science to combat all forms of dogmatism—religious, ideological, and even curricular (as found in the life-adjustment program).

Though the PSSC and BSCS course materials were designed to supplant the functional life-adjustment curriculum of the late 1940s and early 1950s, they were, in many ways, no less functional than what they replaced. To say that the educators had a social agenda and that the scientists did not, that they were only interested in "restoring the primacy of subject matter" to the curriculum, glosses over the fundamental dynamic of this educational reform movement. By laying bare the instrumental nature of various curricular elements at any point in history, as I have tried to do here, one can begin to understand the role of social and political forces in the shaping of school-subject knowledge—to appreciate that curriculum development is always guided by political considerations that substantively affect the nature of instruction. The hope is that we might begin to critically examine the process and materials of education in this light and not continue to think of academic school subjects, especially in the seemingly objective area of the sciences, as simply natural reflections of the academic disciplines they represent. All representations of science in schools embody social and political ends. We would do well to keep this in mind as we undertake the next round of science education reforms in this new century.

# LIST OF ABBREVIATIONS

| | |
|---|---|
| AAAS | American Association for the Advancement of Science |
| *AJP* | *American Journal of Physics* |
| *BAS* | *Bulletin of the Atomic Scientists* |
| *HSPS* | *Historical Studies in the Physical and Biological Sciences* |
| *JHB* | *Journal of the History of Biology* |
| MIT | Massachusetts Institute of Technology |
| NAS | National Academy of Sciences |
| NCDSE | National Committee for the Development of Scientists and Engineers |
| NEA | National Education Association |
| NSC | National Security Council |
| NSF | National Science Foundation |
| *NYT* | *New York Times* |
| SICTSE | Special Interdepartmental Committee on the Training of Scientists and Engineers |
| *USN* | *U.S. News and World Report* |

# LIST OF MANUSCRIPT
# COLLECTIONS ABBREVIATED
# IN NOTES

American Institute of Biological Sciences, Washington, D.C.
    AIBS Papers                                      [AIBS Papers]

American Philosphical Society, Philadelphia, Penn.
    Bentley Glass Papers                       [Glass Papers]

Bentley Historical Library, University of Michigan, Ann Arbor
    Marston Bates Papers                     [Bates Papers]

Biological Sciences Curriculum Study, Colorado Springs, Colorado
    BSCS Papers                                  [BSCS Papers]

Eisenhower Presidential Library
    Eisenhower Papers of the President (Ann Whitman File)
        Cabinet Series                     [AW/CS]
        NSC Series                        [AW/NSC]
    Oveta Culp Hobby Papers              [Hobby Papers]
    Records of the Office of the Special Asst.
    for Sci. and Tech.                      [OSAST]
    Records of the Office of the Staff Secretary
    (White House)                        [WHSS]
    Records of the President's Committee
    on Scientists and Engineers        [PCSE]
    Records of the President's Science
    Advisory Committee                   [PSAC]
    White House Central Files, General File    [WHCF]

Library of Congress
    Alan T. Waterman Papers               [Waterman Papers]

Lilly Library, Indiana University
    Hermann J. Muller Papers            [Muller Papers]
    Tracy Sonneborn Papers              [Sonneborn Papers]

Massachusetts Institute of Technology Archives
    ESI-PSSC-EDC Papers              [PSSC Papers]
    James R. Killian Papers           [Killian Papers]
    Jerrold Zacharias Papers         [Zacharias Papers]
    PSSC Oral History Collection    [PSSC/OHC]
    Records of the Office of the President,
    1930–1958, Compton-Killian      [C/K Papers]

National Academy of Sciences Archives
    Central Policy Files
        Advisory Board on Education    [NAS/ABE]
        Biology and Agriculture/Committee
        on Educational Policies        [NAS/CEP]
        Physical Science/Conf. on the
        Production of Physicists       [NAS/CPP]

National Archives
    Records of the Office of Education, RG 12
        Office Files of the Commissioner
        of Education               [OFCE]
        Selected Central Files         [USOE Sel. Central Files]
        Elliot Richardson NDEA Files   [NDEA Files]
    Records of the Office of Science
    and Technology, RG 359         [OST]
    Records of the National Science
    Foundation, RG 307
        Office of the Director, Subject Files   [NSF/ODSF]
        Assoc. Dir. for International
        and Educational Programs     [NSF Assoc. Dir. Papers]
        Historian File              [NSF/HF]
    Records of the State Department, RG 59
        Lot File 52–283, Project Troy    [Troy Report Files]

Pennsylvania State University Archives
    C. Ray Carpenter Papers        [Carpenter Papers]

Truman Presidential Library
    James E. Webb Papers          [Webb Papers]

University of Illinois, Urbana-Champaign Archives
    Advertising Council Archives     [Ad Council Archives]
    Arthur E. Bestor, Jr. Papers     [Bestor Papers]
    Council for Basic Education Papers  [CBE]

# NOTES

## Introduction

1. Educational Policies Commission, "The Controlling Purposes of American Education," unpublished typescript, 8 January 1960, pp. 3, 7, 9, box 3, Muller Papers. This document was later published under the title *The Central Purpose of American Education* (Washington, D.C.: National Education Association, 1961).

2. See, for example, Educational Policies Commission, *Education for All American Youth* (Washington, D.C.: National Education Association, 1944).

3. See Bruce V. Lewenstein, "'Public Understanding of Science' in America, 1945–1965" (Ph.D. diss., University of Pennsylvania, 1987); and James Gilbert, *Redeeming Culture: American Religion in an Age of Science* (Chicago: University of Chicago Press, 1997).

4. For a survey of the changing nature of school curriculum, see Herbert M. Kliebard, *The Struggle for the American Curriculum, 1893–1958,* 2d ed. (New York: Routledge, 1995). For figures on National Science Foundation expenditures on curriculum development and education programs, see House Committee on Science and Technology, Subcommittee on Science, Research, and Technology, *The National Science Foundation and Pre-College Science Education: 1950–1975,* report prepared by Science Policy Research Division, 94th Cong., 2d sess., 1976, Committee Print. On estimates of usage, see Iris R. Weiss, *Report of the 1977 National Survey of Science, Mathematics, and Social Studies Education* (Washington, D.C.: U.S. Government Printing Office, 1978), 79.

5. Kliebard, *Struggle for the American Curriculum,* xvii.

6. For a summary of the social studies curriculum controversy, see "The National Science Foundation and Pre-College Science Education." On the changing priorities of education in the 1970s, see Diane Ravitch, *The Troubled Crusade: American Education, 1945–1980* (New York: Basic Books, 1983), 228–266; and Joel Spring, *The Sorting Machine Revisited: American Educational Policy Since 1945,* updated ed. (New York: Longman, 1989), 93–150.

7. Two of the more influential statements on science education that have assimilated the intellectual legacy of the NSF curriculum projects are: American Association for the Advancement of Science, *Project 2061: Science for All Americans* (Washington, D.C.: American Association for the Advancement of Science, 1989); and National Research Council, *National Science Education Standards* (Washington, D.C.: National Academy Press, 1996).

8. Zacharias to D. McGee, 16 November 1956, box 15, Zacharias Papers.

## Chapter 1

1. James T. Patterson, *Grand Expectations: The United States, 1945–1974* (New York: Oxford University Press, 1996), 77.

2. Truman quoted in Walter L. Hixson, *Parting the Curtain: Propaganda, Culture, and the Cold War, 1945–1961* (New York: St. Martins Press, 1997), 14.

3. Arthur Zilversmit, *Changing Schools: Progressive Education Theory and Practice, 1930–1960* (Chicago: University of Chicago Press, 1993), 91.

4. Elaine Tyler May, *Homeward Bound: American Families in the Cold War Era* (New York: Basic Books, 1988), 136.

5. Patterson, *Grand Expectations,* 77.

6. Zilversmit, *Changing Schools,* 91; "Nation Tackles School Crisis," *Life,* 26 September 1955, 31–37; Diane Ravitch, *The Troubled Crusade: American Education, 1945–1980* (New York: Basic Books, 1983), 6; "Back to School—and a Bigger Jam," *USN,* 4 September 1953, 22–23.

7. Zilversmit, *Changing Schools,* 90–102; "Classroom Crowding Gets Worse," *USN,* 9 September 1949, 15; "Crisis in Education Persists," *USN,* 8 September 1950, 38–39; "Back to School—and a Bigger Jam;" and "Nation Tackles School Crisis."

8. "Classroom Crowding Gets Worse," 15.

9. "Back to School—and a Bigger Jam," 22–23.

10. Ravitch, *Troubled Crusade,* 26–41; Stephen E. Ambrose, *Eisenhower: Soldier and President* (New York: Simon and Schuster, 1990), 417; Richard L. Strout, "The Big School Scandal," *New Republic,* 19 September 1960, 11–12.

11. Stephen E. Ambrose, *Rise to Globalism: American Foreign Policy Since 1938,* 7th ed. (New York: Penguin Books, 1993), 80.

12. Zilversmit, *Changing Schools,* 103–117; Ravitch, *Troubled Crusade,* 61–80.

13. Ravitch, *Troubled Crusade,* 70.

14. David K. Berninghausen, "A Policy to Preserve Free Public Education," *Harvard Educational Review* 21 (1951): 138–154.

15. Charles H. McCormick, *This Nest of Vipers: McCarthyism and Higher Education in the Mundel Affair, 1951–1952* (Urbana: University of Illinois Press, 1989), 60; Zilversmit, *Changing Schools,* 106.

16. Patterson, *Grand Expectations,* 113, 165–173; Ambrose, *Rise to Globalism,* 85, 95–96; quotation from Paul Boyer, *By the Bomb's Early Light: American Thought and Culture at the Dawn of the Atomic Age* (Chapel Hill: University of North Carolina Press, 1994), 101.

17. Boyer, *Bomb's Early Light,* 336–337; Patterson, *Grand Expectations,* 169.

18. "Italy's Peasants Seize the Land," *Life,* 24 April 1950, 44–46; "Communists Riot in Tokyo," *Life,* 19 June 1950, 131; George Fielding Eliot, "How Prepared Are We if Russia Should Attack?" *Look,* 20 June 1950, 31–35; George Fielding Eliot, "Can We Stop the Reds in Asia?" *Look,* 12 September 1950, 66–73; "Young Communists Rally in East Berlin," *Life,* 20 August 1951, 32; and "Communists in the U.S. a Greater Menace Now," *USN,* 4 September 1953, cover.

19. Ambrose, *Rise to Globalism,* 110–111; Stephen J. Whitfield, *The Culture of the Cold War* (Baltimore: Johns Hopkins University Press, 1991) 1–25; Ellen Schrecker, *Many Are the Crimes: McCarthyism in America* (Boston: Little, Brown and Company, 1998), 154–200.

20. William L. O'Neill, *American High: The Years of Confidence, 1945–1960* (New York: Macmillan, 1986), 160; Patterson, *Grand Expectations,* 194–204.

21. Whitfield, *The Culture of the Cold War,* 9.

22. Ellen W. Schrecker, *No Ivory Tower: McCarthyism and the Universities* (New York: Oxford University Press, 1986), 14, 73.

23. Schrecker, *No Ivory Tower,* 73–74; Ravitch, *Troubled Crusade,* 95–98.

24. Sidney Hook, "Academic Integrity and Academic Freedom," *Commentary* 8 (1949): 337 [emphasis in original], 337, 339.

25. Robert W. Iverson, *The Communists and the Schools* (New York: Harcourt Brace, 1959), 253–254; see also James G. Hershberg, *James B. Conant: Harvard to Hiroshima and the Making of the Nuclear Age* (New York: Knopf, 1993), 451.

26. Quoted in David Hulburd, *This Happened in Pasadena* (New York: Macmillan, 1951), 105–106.

27. Irene Corbally Kuhn, "Your Child Is Their Target," *American Legion Magazine,* June 1952, 18–19, 54–56. For a description of the various anticommunist programs the American Legion sponsored, see McCormick, *This Nest of Vipers,* 47–51.

28. Zilversmit, *Changing Schools,* 114.

29. The Pasadena incident led to the publication of two books: Hulburd, *This Happened in Pasadena;* and Mary Louise Allen, *Education or Indoctrination* (Caldwell, Idaho: Caxton Printers, 1955). Hulburd took the side of the schools, and Allen sided with the conservative critics. On the overall scope of the controversy, see Lawrence A. Cremin, *The Transformation of the School: Progressivism in American Education, 1876–1957* (New York: Random House, 1964), 341.

30. The powerful role of the Red Scare in prompting a reappraisal of American education is documented in Ronald Lora, "Education: Schools as Crucible in Cold War America," in *Reshaping America: Society and Institutions, 1945–1960,* ed. Robert H. Bremner and Gary W. Reichard (Columbus: Ohio State University Press, 1982), 223–260.

31. U.S. Office of Education, *Life Adjustment Education for Every Youth* (Washington, D.C.: U.S. Government Printing Office, 1948), 16; Dorothy Broder, "Life Adjustment Education: An Historical Study of a Program of

the United States Office of Education, 1945–1954" (Ph.D. diss., Teachers College Columbia University, 1977), 57–129.

32. Edward A. Krug, *The Shaping of the American High School,* vol. 2 (Madison: University of Wisconsin Press, 1964–1972), 218.

33. Herbert M. Kliebard, *The Struggle for the American Curriculum: 1893–1958,* 2d ed. (New York: Routledge, 1995), 207.

34. Harl R. Douglass, "Breaking with the Past," in *Education for Life Adjustment: Its Meaning and Implementation* (New York: Ronald Press, 1950), 39–40.

35. Collier quoted in Kliebard, *Struggle for the American Curriculum,* 216.

36. U.S. Office of Education, *Life Adjustment Education for Every Youth,* 8.

37. Kliebard, *Struggle for the American Curriculum,* 77–105, 207. For a discussion of the functional curriculum as the emerging paradigm for general education, see Herbert M. Kliebard, "The Liberal Arts Curriculum and Its Enemies: The Effort to Redefine General Education," in *Forging the American Curriculum: Essays in Curriculum History and Theory* (New York: Routledge, 1992), 27–50.

38. Richard H. Pells, *The Liberal Mind in a Conservative Age: American Intellectuals in the 1940s and 1950s* (New York: Harper and Row, 1985), 187; Barry M. Franklin, *Building the American Community: The School Curriculum and the Search for Social Control* (London: Falmer Press, 1986), 119–137.

39. Douglass, "Breaking with the Past," 43.

40. Quoted in Kliebard, *Struggle for the American Curriculum,* 219.

41. David M. Donahue, "Serving Students, Science, or Society? The Secondary School Physics Curriculum in the United States, 1930–1965," *History of Education Quarterly* 33 (1993): 327.

42. Elliot R. Downing and Veva M. McAtee, *Living Things and You: A Biology Textbook* (Chicago: Lyons and Carnahan, 1940), 14. This tendency to tailor subject matter to the needs of the student and society was evident in the early days of high school biology as well, see Philip J. Pauly, "The Development of High School Biology: New York City, 1900–1925," *Isis* 62 (1991): 662–688. For an overview of this trend in science education, see George E. DeBoer, *A History of Ideas in Science Education: Implications for Practice* (New York: Teachers College Press, 1991), 64–107.

43. Stephen Romine, "Areas and Objectives of Life Adjustment Education," in *Education for Life Adjustment,* ed. Douglass, 47.

44. John Dewey, *Democracy and Education* (New York: Macmillan, 1916), 182.

45. The Cardinal Principles Report is the classic document indicating this shift, NEA, *Cardinal Principles of Secondary Education: A Report of the Commission on the Reorganization of Secondary Education* (Washington, D.C.: U.S. Government Printing Office, 1918).

46. William H. Kilpatrick, "Subject Matter and the Educative Process—III," *Journal of Educational Method* 2 (1923): 368.

47. Robert B. Westbrook, *John Dewey and American Democracy* (Ithaca: Cornell University Press, 1991), 504–505.

48. Kliebard, *Struggle for the American Curriculum,* 75–105; see also, Ellen Condliffe Lagemann, *An Elusive Science: The Troubling History of Education Research* (Chicago: University of Chicago Press, 2000), 80–87.

49. Ellen Herman, *The Romance of American Psychology: Political Culture in the Age of Experts* (Berkeley: University of California Press, 1995), 98. For a broader overview, see James H. Capshew, *Psychologists on the March: Science, Practice, and Professional Identity in America, 1929–1969* (New York: Cambridge University Press, 1999), 39–70; and Peter Buck, "Adjusting to Military Life: The Social Sciences Go to War, 1941–1950," in *Military Enterprise and Technological Change: Perspectives on the American Experience,* ed. M. R. Smith (Cambridge: MIT Press, 1985), 205–252.

50. Herman, *Romance of American Psychology,* 96.

51. W. H. Cowley, "Jabberwocky versus Maturity," in *Trends in Student Personnel Work,* ed. E. G. Williamson (Minneapolis: University of Minnesota Press, 1949), 342–350.

52. Quoted in Herman, *Romance of American Psychology,* 100.

53. Herman, *Romance of American Psychology,* 123. See also, Capshew, *Psychologists on the March,* 116–142.

54. Douglass, "Breaking with the Past," 43.

55. Broder, "Life Adjustment Education," 57.

56. Romine, "Areas and Objectives of Life Adjustment Education," 60–61.

57. Zilversmit, *Changing Schools,* 97–98.

58. U.S. Office of Education, *Life Adjustment Education for Every Youth,* 56.

59. Glenn E. Smith, *Principles and Practices of the Guidance Program* (New York: Macmillan, 1951), 5.

60. Smith, *Principles and Practices of the Guidance Program,* 366.

61. Boyer, *Bomb's Early Light,* 166–177.

62. Cremin, *Transformation of the School,* 344.

63. Arthur E. Bestor, *Educational Wastelands* (Urbana: University of Illinois Press, 1953). Other notable works that contributed to the reexamination of the public schools included Bernard Iddings Bell, *Crisis in Education: A Challenge to American Complacency* (New York: Whittlesey House, 1949); Albert Lynd, *Quackery in the Public Schools* (Boston: Little, Brown, 1953); and Paul Woodring, *Let's Talk Sense about Our Schools* (New York: McGraw-Hill, 1953).

64. Robert H. Beck, "The New Conservatism and the New Humanism," *Teachers College Record* 63 (1962): 435–444. For a more detailed survey of the various arguments made by these critics of education, see Joel Spring, *The Sorting Machine Revisited: National Education Policy since 1945,* Updated ed. (New York: Longman, 1989), 10–24.

65. Raywid describes this series of attacks as the second wave, following the first wave of red-baiting in the schools by reactionary groups, Mary Anne Raywid, *The Ax-Grinders* (New York: Macmillan, 1962).

66. For an extended discussion, see Richard Hofstadter, *Anti-Intellectualism in American Life* (New York: Knopf, 1963), 24–51.

67. Patterson, *Grand Expectations,* 254.

68. This led to the establishment of the Great Books movement, the historical background of which is described in James Sloan Allen, *The Romance of Commerce and Culture: Capitalism, Modernism, and the Chicago-Aspen Crusade for Cultural Reform* (Chicago: University of Chicago Press, 1983).

69. The conflict between the new humanists and the social scientists was an important backdrop to the science curriculum reform movement. The social scientists were aligned, for the most part, with the broader social philosophy of scientific naturalism, also referred to as secular humanism, of which John Dewey was the leading intellectual figure. Individuals of this persuasion advocated the expansion of scientific thinking into all facets of society, including the controversial domain of ethics and morality. The "new humanists" described here were not humanists in this sense of the term, but rather academic traditionalists. Groups opposed to the scientific humanist agenda included these academic traditionalists, who sought to preserve the place of their disciplines in the academy, and religious intellectuals (particularly Catholics), who argued that the moral foundations of society could not be arrived at through science. On the fortunes of scientific naturalism, see Edward A. Purcell, Jr., *The Crisis of Democratic Theory: Scientific Naturalism and the Problem of Value* (Lexington: University Press of Kentucky, 1973), 117–158, 179–217, 235–266. On the pivotal role of Dewey in the debate, see James Gilbert, *Redeeming Culture: American Religion in an Age of Science* (Chicago: University of Chicago Press, 1997), 63–93.

70. Mortimer Smith, *And Madly Teach* (Chicago: Henry Regnery, 1949), 67.

71. Harry J. Fuller, "The Emperor's New Clothes or *Prius Dementat,*" *Scientific Monthly* 72 (1951): 35; Bestor to C. W. Clabaugh, 22 February 1952, box 78, Bestor Papers; H. L. Clapp to M. Smith, 20 December 1949, box 10, CBE.

72. David B. Tyack, *The One Best System: A History of American Urban Education* (Cambridge: Harvard University Press, 1974), 126–176.

73. Bestor, *Educational Wastelands,* 101–121.

74. Bestor, *Educational Wastelands,* 112.

75. Bestor, "The Educators: A Wrecking Crew in Action," unpublished typescript, 1952, box 78, Bestor Papers; for a similar view, see Stewart Scott Cairns, "Mathematics and the Educational Octopus," *Scientific Monthly* 76 (1953): 237.

76. Zilversmit, *Changing Schools,* 114.

77. Marlene M. Wentworth, "From Chautauqua to Wastelands: The Bestors and American Education, 1905–1955" (Ph.D. diss., University of Illinois, Urbana-Champaign, 1992), 273.

78. Quoted in Wentworth, "From Chautauqua to Wastelands," 284.

79. Gustave Albrecht to Bestor, 4 November 1955, box 85, Bestor Papers.

80. The details of this are given in Arthur E. Bestor, *The Restoration of Learning: A Program for Redeeming the Unfulfilled Promise of American Education* (New York: Knopf, 1956), n1, 224–225.

81. Bestor to A. M. Schlesinger, 6 December 1952, box 79, Bestor Papers.

82. Commenting on various reactions to his criticisms, he wrote, "I am sorry to be now so peremptorily excommunicated and to be vilified as 'unde-mocratic and un-American' in anonymous communications in the public press." Bestor to Kenneth Benne, 14 March 1952, box 78, Bestor Papers.

83. Wentworth, "From Chautauqua to Wastelands," 351–352.

84. Bestor, *Educational Wastelands*, 26, 34.

85. Quoted in Wentworth, "From Chautauqua to Wastelands," 315–316.

86. Bestor, *Educational Wastelands*, 7.

87. Bestor, *Educational Wastelands*, 13.

88. Fuller, "Emperor's New Clothes," 38. Fuller's observations about the nature of biology teaching were supported by a study undertaken by Oscar Rid-dle and his colleagues in the early 1940s, Oscar Riddle, ed., *The Teaching of Biology in Secondary Schools of the United States* (Evanston: Committee on the Teaching of Biology of the Union of American Biological Societies, 1942).

89. Bestor, *Educational Wastelands*, 36, 18–19.

90. Bestor, *Educational Wastelands*, 55.

91. Smith, *And Madly Teach,* 220; Bestor, *Educational Wastelands*, 22.

92. Bestor, *Educational Wastelands*, 186.

93. Bestor, *Educational Wastelands*, 186, 121.

94. Charles R. Foster, *Psychology for Life Adjustment* (Chicago: American Tech-nical Society, 1951), 417.

95. Quoted in Kliebard, *Struggle for the American Curriculum,* 223.

96. Kliebard, *Struggle for the American Curriculum,* 223.

97. An officer of the Sears-Roebuck Foundation considering a funding re-quest for continued work in the Life Adjustment program commented that the name "'life adjustment' had a rather leftish implication," Broder, "Life Adjustment Education," 52.

98. Patterson, *Grand Expectations,* 65.

99. For example, Nolan C. Kearny, "Educating the Gifted Child," *New Repub-lic,* 27 March 1957, 13–15; Anita Bell Morgan, "Identification and Guid-ance of Gifted Children," *Scientific Monthly* 60 (1955): 171–174; Robert C. Wilson, "The Under-Educated: How We Have Neglected the Bright Child," *Atlantic Monthly,* May 1955, 60–62; Eugene Youngert, "Giving the Bright Student a Break," *Atlantic Monthly,* June 1956, 39–41; Stuart Chase, "Five Year Plan for Eggheads," *Progressive,* July 1956, 5–6; Arthur Bestor, "Educating the Gifted Child," *New Republic,* 4 March 1957, 12–16.

## Chapter 2

1. Arthur E. Bestor, *Educational Wastelands* (Urbana: University of Illinois Press, 1953), 197–206.

2. Council for Basic Education, meeting minutes, 30 November 1954, box 83, Bestor Papers.

3. Arthur E. Bestor, "Aimlessness in Education," *Scientific Monthly* 75 (1952): 109–116; Harry J. Fuller, "The Emperor's New Clothes or *Prius Dementat*," *Scientific Monthly* 72 (1951): 32–41.

4. Lynn White to Bestor, 17 December 1952, box 79, Bestor Papers.

5. A discussion of the various public images of science and scientists for the first half of the twentieth century can be found in Marcel C. LaFollete, *Making Science Our Own: Public Images of Science, 1910–1955* (Chicago: University of Chicago Press, 1990). On the public authority of science see Thomas F. Gieryn, *Cultural Boundaries of Science: Credibility on the Line* (Chicago: University of Chicago Press, 1999), 1–35.

6. S. S. Schweber, "Big Science in Context: Cornell and MIT," in *Big Science: The Growth of Large-Scale Research,* ed. Peter Galison and Bruce Hevly (Stanford: Stanford University Press, 1992), 156; Daniel J. Kevles, *The Physicists: The History of a Scientific Community in Modern America* (Cambridge: Harvard University Press, 1987), 287–323; Walter A. McDougall, *The Heavens and the Earth: A Political History of the Space Age* (New York: Basic Books, 1985), 5–6.

7. Kevles, *The Physicists,* 302–323; Dexter Masters, "We Outsmarted Them on Radar," *Saturday Evening Post,* 8 September 1945, 20; David P. Adams, "The Penicillin Mystique and the Popular Press (1935–1950)," *Pharmacy in History* 26 (1984): 134–142.

8. Donald Porter Geddes, *The Atomic Age Opens* (New York: Pocket Books, 1945), 33.

9. Kevles, *The Physicists,* 324–334; Paul Boyer, *By the Bomb's Early Light: American Thought and Culture at the Dawn of the Atomic Age* (Chapel Hill: University of North Carolina Press, 1994); U.S. Strategic Bombing Survey, *The Effects of Atomic Bombs on Hiroshima and Nagasaki* (Washington, D.C.: U.S. Government Printing Office, 1946).

10. Boyer, *Bomb's Early Light,* 60.

11. Francis Sill Wickware, "Manhattan Project," *Life,* 20 August 1945, 100.

12. Military technology had already exhibited its debt to science, and spectacular contributions were also made in the medical field with the development of antibiotics and later the polio vaccine. The promise of a war on cancer further contributed to the belief that there was little science could not conquer; see James T. Patterson, *The Dread Disease: Cancer and Modern American Culture* (Cambridge: Harvard University Press, 1987), 137–170. Reference to the "age of science" was growing quite common during the 1950s. The philosopher Herbert Feigl commented in 1955 that "It has been stated and repeated—perhaps *ad nauseum*—that ours is an age of science"; see Feigl, "Aims of Education for Our Age of Science: Reflections of a Logical Empiricist," in *Modern Philosophies and Education: The Fifty-Fourth Yearbook of the National Society for the Study of Education,* ed. Nelson B. Henry (Chicago: University of Chicago Press, 1955), 306. On the "gee-whiz" popularization of science, see John C. Burnham, *How Superstition*

*Won and Science Lost: Popularizing Science and Health in the United States* (New Brunswick: Rutgers University Press, 1987).

13. Kevles, *The Physicists,* 376.

14. Joel H. Hildebrand, "The Professor and His Public," *BAS* 9 (1953): 23–25; Kevles, *The Physicists,* 375–376.

15. Boyer, *Bomb's Early Light,* 334.

16. Paul Forman, "Social Niche and Self-Image of the American Physicist," in *Proceedings of the International Conference on the Restructuring of Physical Sciences in Europe and the United States, 1945–1960,* ed. Michelangelo De Maria, Mario Grilli, and Fabio Sebastiani (Singapore: World Scientific, 1989), 102–104.

17. Daniel J. Kevles, "K1S2: Korea, Science, and the State," in *Big Science,* ed. Galison and Hevly, 312–333.

18. McDougall, *Heavens and the Earth,* 93–96; Stephen E. Ambrose, *Rise to Globalism: American Foreign Policy Since 1938,* 7th ed. (New York: Penguin Books, 1993), 110–113.

19. Kevles, "K1S2," 325.

20. Quoted in McDougall, *Heavens and the Earth,* 114. The economic rationale for the New Look is described in Stephen E. Ambrose, *Eisenhower: Soldier and President* (New York: Simon and Schuster, 1990), 356.

21. Kevles, "K1S2," 330.

22. Silvan S. Schweber, "The Mutual Embrace of Science and the Military: ONR and the Growth of Physics in the United States after World War II," in *Science, Technology, and the Military,* ed. Everett Mendelsohn, Merit Roe Smith, and Peter Weingart (Dordrecht: Kluwer, 1988), 3–45; see also Roger L. Geiger, "Science, Universities, and National Defense, 1945–1970," *Osiris,* 2d series, 7 (1992): 26–48; and Kenneth M. Jones, "The Government-Science Complex," in *Reshaping America: Society and Institutions, 1945–1960,* ed. Robert H. Bremner and Gary W. Reichard (Columbus: Ohio State University Press, 1982), 315–342. The quotation is from McDougall, *Heavens and the Earth,* 5. Though McDougall places the arrival of technocracy in the late 1950s with the launch of *Sputnik,* its origins can be traced to the events of 1949, where most historians of science locate its emergence; see, for example, Kevles, "K1S2," 312–333.

23. Stuart W. Leslie, *The Cold War and American Science: The Military-Industrial-Academic Complex at MIT and Stanford* (New York: Columbia University Press, 1993), 2.

24. Peter Galison, "Physics between War and Peace," in *Science, Technology, and the Military,* ed. Mendelsohn, et al., 47–86.

25. Kevles, *The Physicists,* 341, 367.

26. Leslie, *Cold War and American Science,* 44–75; Kevles, "K1S2," 329.

27. Philip Morrison, "The Laboratory Demobilizes . . ." *BAS* 2 (1946): 1–2. For discussion of the various sources of the ample nondirected government funding, see Schweber, "Mutual Embrace," 3–45; and Geiger, "Science, Universities, and National Defense, 1945–1970," 26–48. For further

examination of scientist opposition to military patronage, see Jessica Wang, *American Science in an Age of Anxiety: Scientists, Anticommunism, and the Cold War* (Chapel Hill: University of North Carolina Press, 1999), 38–42.

28. David A. Hollinger, "Free Enterprise and Free Inquiry: The Emergence of Laissez-Faire Communitarianism in the Ideology of Science in the United States," *New Literary History* 21 (1990): 900.

29. Mark Silk, *Spiritual Politics: Religion and America Since World War II* (New York: Simon and Schuster, 1988), 38, 63–64, 106, 95–100. The increasing level of institutional religiosity was not necessarily accompanied by a significant deepening of spiritualism in the country. This fact was particularly distressing to intellectuals during the 1950s; James T. Patterson, *Grand Expectations: The United States, 1945–1974* (New York: Oxford University Press, 1996), 332. On the threat to intellectuals, see Stephen P. Weldon, "The Humanist Enterprise from John Dewey to Carl Sagan: A Study of Science and Religion in American Culture" (Ph.D. diss., University of Wisconsin-Madison, 1997), 112–143.

30. For details, see Donald R. McNeil, *The Fight for Fluoridation* (New York: Oxford University Press, 1957).

31. Bernard Mausner and Judith Mausner, "A Study of the Anti-Scientific Attitude," *Scientific American* 189 (1955): 35–39.

32. Quoted in McNeil, *Fight for Fluoridation,* 172.

33. Robert A. Divine, *Blowing on the Wind: The Nuclear Test Ban Debate, 1954–1960* (New York: Oxford University Press, 1978), 4–18.

34. R. B. Lindsay, "The Survival of Physical Science," *Scientific Monthly* 74 (1952): 140; Charles A. Metzner and Judith B. Kessler, "What are the People Thinking?" *BAS* 7 (1951): 341.

35. Quoted in Kevles, *The Physicists,* 346.

36. Daniel J. Kevles, "The National Science Foundation and the Debate over Postwar Research Policy," *Isis* 68 (1977): 5–26.

37. Kevles describes this as the "linear model" used by scientists to justify funding of basic research, Daniel J. Kevles, "R&D and the Arms Race: An Analytical Look," in *Science, Technology, and the Military,* ed. Mendelsohn, et al., 47–86.

38. U.S., Office of Scientific Research and Development (OSRD), *Science—The Endless Frontier: A Report to the President on a Program for Postwar Scientific Research, by Vannevar Bush, 1945* (Washington, D.C.: National Science Foundation, 1960), 13.

39. OSRD, *Science—The Endless Frontier,* 13; and NSF, *Basic Research: A National Resource* (Washington, D.C.: U.S. Government Printing Office, 1957); see also Daniel Lee Kleinman and Mark Solovey, "Hot Science/Cold War: The National Science Foundation after World War Two," *Radical History Review* 63 (1995): 110–139.

40. OSRD, *Science—The Endless Frontier,* 13–14.

41. Kevles, *The Physicists,* 349.

42. Kevles, *The Physicists,* 364–366. Kleinman describes the nature and composition of this elite in Daniel Lee Kleinman, *Politics on the Endless Frontier* (Durham: Duke University Press, 1995), 52–73. On Waterman's background, see J. Merton England, *A Patron for Pure Science: The National Science Foundation's Formative Years, 1945–57* (Washington, D.C.: National Science Foundation, 1982), 118–127.

43. Quoted in James G. Hershberg, *James B. Conant: Harvard to Hiroshima and the Making of the Nuclear Age* (New York: Knopf, 1993), 565.

44. England, *Patron for Pure Science,* 153.

45. Hershberg, *James B. Conant,* 565.

46. England, *Patron for Pure Science,* 160.

47. Wilson quoted in Kevles, *The Physicists,* 383.

48. Rabinowitch quoted in England, *Patron for Pure Science,* 157.

49. Bush to Bronk, 1 September 1953, box 12, NSF/ODSF.

50. McDougall, *Heavens and the Earth,* 72.

51. See Edward A. Purcell, Jr., *The Crisis of Democratic Theory: Scientific Naturalism and the Problem of Value* (Lexington: University Press of Kentucky, 1973), 117–266.

52. For overview, see Wang, *American Science in an Age of Anxiety;* Ellen W. Schrecker, *No Ivory Tower: McCarthyism and the Universities* (New York: Oxford University Press, 1986), 126–160; and Kevles, *The Physicists,* 378–384. For a contemporary appraisal of the impact of the Red Scare on the scientific community, see Edward A. Shils, *The Torment of Secrecy: The Background and Consequences of American Security Policies* (Glencoe, IL: Free Press, 1956).

53. Schrecker, *No Ivory Tower,* 131–132; Kevles, *The Physicists,* 330.

54. M. Stanley Livingston, "The Scientist and Security," *Current History* 29 (1955): 219. Similar concerns were expressed by NAS president Detlev Bronk; William L. Laurence, "Bronk Says Fear Cripples Science," *NYT,* 28 April 1954. A comprehensive statement of the problem of science and security was issued by the AAAS in the fall of 1954, William L. Laurence, "Scientists Set Security Goal of Progress, Not Secrecy," *NYT,* 12 December 1954.

55. Schrecker, *No Ivory Tower,* 131.

56. Edward A. Shils, *Torment of Secrecy,* 11; see also, Eugene Rabinowitch, "The Atomic Bomb Secret—Fifteen Years Later," *BAS* 22 (December 1966): 2–3, 25.

57. Lloyd V. Berkner, "Science and National Strength," *Physics Today* 6 (1953): 9.

58. Warren Weaver to Hubert H. Humphrey, 15 February 1957, NSF/HF. For more on this myth, see Schrecker, *No Ivory Tower,* 132; Robert W. Iversen, *The Communists and the Schools* (New York: Harcourt Brace and Co., 1959), 296; and Eugene Rabinowitch, "The Atomic Secrets," *BAS* 3 (1947): 33, 68.

59. "Fight for Mind's Freedom," *Science News Letter* 61 (1952): 22–23; Melba Phillips, "Dangers Confronting American Science," *Science* 116 (1952): 439–443.

60. Ernest C. Pollard and John B. Phelps, "Science and Security," *Progressive,* November 1954, 10.

61. Karl T. Compton, "Science and Security," *BAS* 4 (1948): 375.

62. Wang, *American Science in an Age of Anxiety,* 198–199.

63. For concern in the academic community, see Alonzo F. Myers, "Thought Control: U.S. Style," *Progressive,* September 1951, 9–11.

64. "Fight for Mind's Freedom," 23.

65. For detailed account, see David Halberstam, *The Fifties* (New York: Villard Books, 1993), 330–335, 342–345, 349–354; and Jeff Broadwater, *Eisenhower and the Anti-Communist Crusade* (Chapel Hill: University of North Carolina Press, 1992), 96–101. For the reaction of the intellectual community, see Richard H. Pells, *The Liberal Mind in a Conservative Age: American Intellectuals in the 1940s and 1950s* (New York: Harper and Row, 1985), 344.

66. Hershberg, *James B. Conant,* 681.

67. "The Scientist in Society," *Physics Today* 7 (1954): 7.

68. Vannevar Bush, "If We Alienate Our Scientists," *New York Times Magazine,* 13 June 1954, 9, 71.

69. Bethe to Bush, 18 June 1954, box 138, Vannevar Bush Papers, Library of Congress.

70. Maurice B. Visscher, "Scientists in a Mad World," *Nation,* 24 January 1953, 69.

71. Visscher, "Scientists in a Mad World," 70.

72. Eugene Rabinowitch quoted in Kevles, *The Physicists,* 384.

73. V. F. Weisskopf, "Science for Its Own Sake," *Scientific Monthly* 78 (1954): 133.

74. Weisskopf, "Science for Its Own Sake," 133; a common component of this view of science was the determination and dissemination of the truth as discussed by L. J. F. Brimble, "The Exposition of Truth," *BAS* 4 (1948): 141–144; and James Franck, "The Social Task of the Scientist," *BAS* 3 (1947): 70.

75. Phillips, "Dangers Confronting American Science," 439–440.

76. Alan T. Waterman, "Acceptance of Science," *Scientific Monthly* 80 (1955): 13–14.

77. Bruce V. Lewenstein, "'Public Understanding of Science' in America, 1945–1965" (Ph.D. diss., University of Pennsylvania, 1987), 191–192.

78. Marston Bates to AAAS, 7 September 1951, box 4, Bates Papers.

79. Conant quoted in Hershberg, *James B. Conant,* 409.

80. James B. Conant, *The Growth of the Experimental Sciences: An Experiment in General Education* (Cambridge: Harvard University press, 1949), 2.

81. See Peter S. Buck and Barbara Gutmann Rosenkrantz, "The Worm in the Core: Science and General Education," in *Transformation and Tradition in the Sciences: Essays in Honor of I. Bernard Cohen,* ed. Everett Mendelsohn (New York: Cambridge University Press, 1984), 371–394; and I. Bernard Cohen and Fletcher G. Watson, eds., *General Education in Science* (Cambridge: Harvard University Press, 1952); Earl J. McGrath, "The General Education

Movement," *Journal of General Education* 1 (1946): 3–8. On the goals of the Harvard courses, see Steve Fuller, *Thomas Kuhn: A Philosophical History for Our Times* (Chicago: University of Chicago Press, 2000), 150–221.

82. A. J. Carlson, A. C. Ivy, and Ralph A. Rohweder to Howard Meyerhoff, 7 June 1951, box 1, NSF/ODSF.

83. "AAAS Policy," *Scientific Monthly* 73 (1951): 335–336. For a detailed discussion, see Sally Gregory Kohlstedt, Michael M. Sokal, and Bruce V. Lewenstein, *The Establishment of Science in America: 150 Years of the American Association for the Advancement of Science* (New Brunswick: Rutgers University Press, 1999), 109–112.

84. Warren Weaver to Waterman, 8 January 1953, box 1, NSF/ODSF.

85. Dael Wolfle, *Renewing a Scientific Society: The American Association for the Advancement of Science from World War II to 1970* (Washington, D.C.: American Association for the Advancement of Science, 1989), 189–209; for details, see also Lewenstein, "Public Understanding of Science," 173–187.

86. Edward Shils, "The Scientific Community: Thoughts After Hamburg," *BAS* 10 (1954): 151.

87. See Samuel K. Allison, "Loyalty, Security and Scientific Research in the United States," in *Proceedings of the Hamburg Congress on Science and Freedom, Congress for Cultural Freedom* (London: Martin Secker & Warburg Ltd., 1955), 78–86.

88. For a summary, see Shils, "Scientific Community," 151–155.

89. Henry E. Sigerist, "Science and Democracy," *Science and Society* 2 (1938): 297.

90. J. D. Bernal, "Dialectical Materialism and Modern Science," *Science and Society* 2 (1938): 63.

91. Loren R. Graham, *Science in Russia and the Soviet Union: A Short History* (New York: Cambridge University Press, 1993), 93–103; McDougall, *Heavens and the Earth,* 23–33.

92. Hermann J. Muller, "The Crushing of Genetics in the USSR," *BAS* 4 (1948): 369; see also Theodosius Dobzhansky, "The Fate of Biological Science in Russia," in *Proceedings of the Hamburg Congress,* 212–223.

93. Shils, "Scientific Community," 151.

94. Shils, "Scientific Community," 154; Purcell, *Crisis of Democratic Theory,* 235–266. Contrasting the nature of science with various other knowledge systems became a fairly common intellectual exercise among philosophers and scientists, see, for example, A. Cornelius Benjamin, "Science and its Presuppositions," *Scientific Monthly* 73 (1951): 150–153; Adolf Grünbaum, "Science and Ideology," *Scientific Monthly* 79 (1954): 13–19; Charles Hartshorne, "Mind, Matter, and Freedom," *Scientific Monthly* 78 (1954): 314–320; Ira Jewell Williams, "Science and Dogma," *BAS* 9 (1953): 219–220; Lawrence Moran, "Catholicism, Communism, and Science," *BAS* 9 (1953): 49–64; and Sidney Hook, "Science and Dialectical Materialism," 182–195, and Bruno Snell, "Science and Dogma," 134–140, both in *Proceedings of the Hamburg Congress.*

95. See David A. Hollinger, "The Defense of Democracy and Robert K. Merton's Formulation of the Scientific Ethos," *Knowledge and Society* 4 (1983): 1–15. For a broader survey of the historical relationship between science and democracy, see Roy Macleod, "Science and Democracy: Historical Reflections on Present Discontents," *Minerva* 35 (1997): 369–384; and Jessica Wang, "Merton's Shadow: Perspectives on Science and Democracy since 1940," *HSPS* 30 (1999): 279–306.

96. Dewey quoted in Purcell, *Crisis of Democratic Theory*, 206.

97. David Thomson, "Science and Democracy," *Nation*, 10 September 1955, 224.

98. David A. Hollinger, "Science as a Weapon in *Kulturkämpfe* in the United States During and After World War II," *Isis* 86 (1995): 327.

99. Shils, "Scientific Community," 151, 154.

100. See, for example, George R. Harrison, "Faith and the Scientist," *Atlantic Monthly*, December 1953, 48–53; and Vannevar Bush, "For Man to Know," *Atlantic Monthly*, August 1955, 29–34.

101. Hollinger, "Free Enterprise and Free Inquiry," 911.

102. Shils, Torment of Secrecy, 9.

## Chapter 3

1. Earl J. McGrath to Rufus Miles, 8 January 1953, box 1, OFCE.

2. Philip N. Powers, "The Changing Manpower Picture," *Scientific Monthly* 70 (1950): 170.

3. David F. Noble, *Forces of Production: A Social History of Industrial Automation* (New York: Oxford University Press, 1984), 5–9. On the concern with scientific manpower shortages in physics after the war, see David Kaiser, "Making Theory: Physics and Physicists in Postwar America" (Ph.D. diss., Harvard University, 2000), 12–56.

4. "Statement Regarding the Scientific and Engineering Manpower Problem in the US," NSF memo, 22 August 1956, box 22, PCSE.

5. Margaret W. Rossiter, *Women Scientists in America: Before Affirmative Action, 1940–1972* (Baltimore: Johns Hopkins University Press, 1995), 52. Lewenstein describes the readership of *Scientific American* as consisting nearly exclusively of the scientific/technocratic elite, Bruce V. Lewenstein, "The Meaning of 'Public Understanding of Science' in the United States after World War II," *Public Understanding of Science* 1 (1992): 51–52.

6. Howard A. Meyerhoff, "Arms and Manpower," *BAS* 10 (1954): 119.

7. For a discussion of the fellowship program, see J. Merton England, *A Patron for Pure Science: The National Science Foundation's Formative Years, 1945–57* (Washington, D.C.: National Science Foundation, 1982), 228–235. On draft deferments, see Joel Spring, *The Sorting Machine Revisited: American Educational Policy Since 1945,* updated ed. (New York: Longman, 1989), 53. The special issue was *Scientific American*, 185, no. 3 (1951).

8. "Scientific Womanpower," *BAS* 7 (1951): 34.

9. Arthur S. Flemming, "The Human Resources of the U.S.: Mobilization," *Scientific American* 185 (1951): 94.

10. "Scientific Womanpower," 34.

11. For detailed discussion of the suggested roles women might play during the crisis, see Rossiter, *Women Scientists in America,* 50–57.

12. Quoted in Hillier Krieghbaum and Hugh Rawson, *An Investment in Knowledge: The First Dozen Years of the National Science Foundation's Summer Institutes Programs to Improve Secondary School Science and Mathematics Teaching, 1954–1965* (New York: New York University Press, 1969), 66.

13. The most complete account of the NSF summer institutes can be found in Krieghbaum and Rawson, *Investment in Knowledge,* 65–82.

14. Quoted in Krieghbaum and Rawson, *Investment in Knowledge,* 66.

15. Two of the more well-known reports were the Barton Report, issued by the American Institute of Physics, and the Trytten Report, which was submitted to the director of the Selective Service System by M. H. Trytten in 1950; see "The Organization of Scientific Manpower: Reports of Government Advisory Groups," *BAS* 6 (1950): 355, 375–378. On the intensification of the crisis due to the Korean War, see M. H. Trytten, "The Human Resources of the U.S.: Scientists," *Scientific American* 185 (1951): 72; see also, Leonard I. Katzin, "Scientific Manpower," *BAS* 8 (1952): 27–30. Early appeals to the education community were made by Dael Wolfle and Maynard Boring; Wolfle, "America's Intellectual Resources," 125–135; and Boring, "Supply and Demand for Engineers," 136–139, both in the *Bulletin of the National Association of Secondary-School Principals* 36 (1952).

16. National Manpower Council, *A Report on the National Manpower Council* (New York: Graduate School of Business, Columbia University, 1954) 7.

17. National Manpower Council, *A Policy for Scientific and Professional Manpower* (New York: Columbia University Press, 1953).

18. National Manpower Council, *Report on the National Manpower Council,* 25.

19. Spring, *Sorting Machine,* 54–55.

20. John Lewis Gaddis, *Strategies of Containment: A Critical Appraisal of Postwar American National Security Policy* (New York: Oxford University Press, 1982), 127–163.

21. "Soviet Scientific and Technical Manpower," cabinet paper included in 30 April 1954 cabinet meeting, box 3, AW/CS.

22. NSC meeting minutes, 22 October 1953, box 4, AW/NSC; cabinet meeting agenda and minutes, 30 April 1954, box 3, AW/CS.

23. Arthur Flemming to Waterman, 24 May 1954, box 18, NSF/ODSF.

24. By January of 1955, the expansion of scientific training became a formal component of national security policy of the Eisenhower administration. See NSC 5501, Basic National Security Policy, 6 January 1955, discussed on 5 January 1955, NSC meeting minutes, box 6, AW/NSC.

25. News conference, 10 June 1954, Dwight D. Eisenhower, *Public Papers of the Presidents of the United States: Dwight D. Eisenhower, 1953–1961,* vol. 2

(Washington, D.C.: U.S. Government Printing Office, 1960), 549; Elie Abel, "Eisenhower Calls Red Peril Global and Truly Serious," *NYT,* 11 June 1954.

26. For an account of the threat the interdepartmental committee posed to NSF, see England, *Patron for Pure Science,* 248–254; on Kelly's background, 133.

27. Flemming to Waterman, 24 May 1954, box 18, NSF/ODSF.

28. Alan T. Waterman (ATW) diary note, 21 October 1954, box 5, NSF/ODSF.

29. ATW diary notes, 21 July 1954, 21 October 1954, box 5; Waterman to Flemming, 30 November 1954, box 18; NSF Notes for SICTSE meeting, 30 November 1954, box 18, NSF/ODSF; NSF Scientific Personnel and Education (SPE) Divisional Committee meeting minutes, 10 November 1954, NSF/HF.

30. SPE Divisional Committee meeting minutes, 19 May 1954, NSF/HF.

31. ATW diary note, 21 July 1954, box 5, NSF/ODSF.

32. National Manpower Council, *Policy for Scientific and Professional Manpower,* 199–218. With respect to physics education, see "NRC/AIP Conference on the Production of Physicists," *Physics Today* 8 (1955): 6–12. SPE Divisional Committee meeting minutes, 19 May 1954, NSF/HF.

33. National Manpower Council, *Policy for Scientific and Professional Manpower,* 213–217.

34. ATW diary note, 21 July 1954, box 5, NSF/ODSF.

35. These conditions were reported in a series of articles on the postwar education crisis by Benjamin Fine in 1947, see Arthur Zilversmit, *Changing Schools: Progressive Education Theory and Practice, 1930–1960* (Chicago: University of Chicago Press, 1993), 92–93.

36. National Manpower Council, *Policy for Scientific and Professional Manpower,* 201–206.

37. National Manpower Council, *Policy for Scientific and Professional Manpower,* 206.

38. SPE Divisional Committee meeting minutes, 22 September 1955, NSF/HF.

39. Leonard O. Olsen and Rollin W. Waite, "Effectiveness of the Case-General Electric Science Fellowship Program for High School Physics Teachers," *AJP* 23 (1955): 424.

40. ATW diary note, 21 October 1954, box 5, NSF/ODSF.

41. The difficulties posed by the professional education bureaucracy was discussed at length in SPE Divisional Committee meeting minutes, 22 September 1955, NSF/HF; as well as by the Educational Advisory Board of the NAS, meeting minutes of Educational Advisory Board/National Research Council, 14 June 1956, box 21, PCSE. That educators represented a serious impediment to educational improvement was a conclusion reached by SICTSE in their final report, "The Development of Scientists and Engineers," April 1955, box 1, PCSE. See also Harold E. Wise, "Critique of Teaching Objectives in Secondary Schools," *AJP* 23 (1955): 415–423.

42. Kelly to Waterman, 3 July 1952, box 18, NSF/ODSF.

43. Krieghbaum and Rawson, *Investment in Knowledge,* 91–104, 129–140.

44. SPE Divisional Committee meeting minutes, 19 May 1954, NSF/HF.

45. Kelly interview, PSSC/OHC.

46. SPE Divisional Committee meeting minutes, 19 May 1954, NSF/HF.

47. ATW diary note, 25 October 1954, NSF/HF.

48. Clarence J. Brown quoted in Mark Solovey, "The Politics of Intellectual Identity and American Social Science, 1945–1970" (Ph.D. diss., University of Wisconsin-Madison, 1996), 91.

49. Ellen Herman, *The Romance of American Psychology: Political Culture in the Age of Experts* (Berkeley: University of California Press, 1995), 195–199; Solovey, "Politics of Intellectual Identity," 105–166; and James T. Patterson, *Brown v. Board of Education: A Civil Rights Milestone and Its Troubled Legacy* (New York: Oxford University Press, 2001), 61–63.

50. ATW diary note, 21 October 1954, box 5, NSF/ODSF; SICTSE, "Report to the President," December 1954, box 1, PCSE; SICTSE, "The Development of Scientists and Engineers," April 1955, box 1, PCSE; SPE Divisional Committee meeting minutes, 10 November 1954, NSF/HF.

51. Kelly to Harry Winne, 12 November 1954, box 18, NSF/ODSF; see also discussion at SPE Divisional Committee meeting, 10 November 1954, NSF/HF.

52. SPE Divisional Committee meeting minutes, 10 November 1954, NSF/HF.

53. Kelly to Harry Winne, 12 November 1954, box 18, NSF/ODSF.

54. National Manpower Council, *Policy for Scientific and Professional Manpower,* 255.

55. Brownell, "Educational Aspects of the Scientific and Technical Manpower Problem," Office of Education Report, 11 May 1956, box 72, OFCE.

56. This was also the conclusion of a group of educators and scientists who met at Cornell University to consider the state of science education in relation to the manpower problem. The organizations represented included the NEA, the National Science Teachers Association, the AAAS, and Westinghouse Electric, "A Proposed Plan for Action Relating to Scientific and Engineering Manpower Development," 9 September 1954, box 72, OFCE.

57. SICTSE, "The Development of Scientists and Engineers," April 1955, box 1, PCSE.

58. Frank J. Munger and Richard F. Fenno, *National Politics and Federal Aid to Education* (Syracuse, NY: Syracuse University Press, 1962), 8–18, 46–47, 65–75; Diane Ravitch, *The Troubled Crusade: American Education, 1945–1980* (New York: Basic Books, 1983), 120; see also Carl F. Kaestle and Marshall S. Smith, "The Federal Role in Elementary and Secondary Education, 1940–1980," *Harvard Educational Review* 52 (1982): 384–408. For the Nixon quotation, cabinet meeting minutes, 14 January 1955, box 4, AW/CS.

59. There apparently was no public demand for federal involvement in local school affairs. In fact, the special interdepartmental committee report had

concluded that the public was "not sufficiently aware of the nature and seriousness of the ... shortage of qualified scientists and engineers;"
SICTSE, "The Development of Scientists and Engineers," p. 16, April
1955, box 1, PCSE.

60. House Committee on Appropriations, *Hearings before the Subcommittee on
Independent Offices,* 84th Cong., 1st sess., 9 February 1955, 234–235, emphasis added.

61. SPE Divisional Committee meeting minutes, 18 May 1955, NSF/HF.

62. Benjamin Fine, "Russia is Overtaking US in Training of Technicians,"
*NYT,* 7 November 1954; "Critical Shortage of Technically Trained Americans is Causing Much Concern," *NYT,* 5 June 1955; "Industry Raises Cry
for More Engineers," *NYT,* 16 October 1955.

63. Fine, "Russia is Overtaking US in Training of Technicians."

64. Fine, "Critical Shortage;" see also Gene Currivan, "Leaders Deplore Lag in
Scientists," *NYT,* 20 November 1956.

65. NSC meeting minutes, 13 October 1955, box 7, AW/NSC.

66. Committee for the White House Conference on Education, *Report of the
White House Conference on Education* (Washington, D.C.: U.S. Government
Printing Office, 1956).

67. Michael Straight, "Schools: US Emergency," *New Republic,* 12 December
1955, 6.

68. "Critics Assured on School Study," *NYT,* 29 November 1955.

69. On Waterman contact, see ATW diary note, 2 May 1955, box 5,
NSF/ODSF; on those solicited for feedback, see "Suggestions for Critics
of Position Statements," 1 May 1955, box 237, C/K Papers.

70. James R. Killian, meeting minutes of the subcommittee "What Should
Our Schools Accomplish," 10 March 1955, box 19, Hobby Papers; Arthur
Bestor to Killian, 1 July 1955; Harry Fuller to Killian, 3 June 1955, box
237, C/K Papers.

71. James R. Killian, "What Should Our Schools Accomplish?" discussion
draft, 20 May 1955, box 237, C/K Papers.

72. Bestor to Killian, 1 July 1955, box 237, C/K Papers.

73. "Critics Assured on School Study"; Hobby also urged conference participants to consider the manpower problem in the advance materials she sent
out, "Suggestions for the White House Conference on Education," December (?) 1954, box 19, Hobby Papers.

74. Killian, "What Should Our Schools Accomplish?" discussion draft, 20 May
1955, box 237, C/K Papers. Many of the various state and regional meetings, which took place in advance of the national meeting, came to the
same conclusions regarding the desirability of a life-adjustment type of
curriculum, "Recommendations of Conference Discussion Groups," New
York City Regional Meeting for the White House Conference on Education, 25 June 1955, box 1; "Summary Report of the Study Committee,"
Wisconsin Conference on Education Regional Conferences, box 9; "What

Should Our Schools Accomplish?" South Carolina Regional Report, box 4, Reports from the White House Conference on Education, 1955–57, OFCE.

75. Straight, "Schools: US Emergency," 7–8.

76. Waterman to National Science Board, 1 January 1956, box 18, NSF/ODSF.

77. England, *Patron for Pure Science*, 219.

78. Nicholas DeWitt, *Soviet Professional Manpower* (Washington, D.C.: National Science Foundation, 1955), 2, 4, 41, 42, 39.

79. Detonation of the deliverable device occurred in November 1955. The Soviet Union's first successful explosion of a stationary hydrogen bomb occurred in August 1953; Walter A. McDougall, *The Heavens and the Earth: A Political History of the Space Age* (New York: Basic Books, 1985), 55.

80. House Committee on Appropriations, *Hearings before the Subcommittee on Independent Offices,* 84th Cong., 2nd sess., 30 January 1956, 522. The level of concern on the part of legislators was reflected in the large number of requests for DeWitt's book from various congressional offices, ATW diary note, 24 February 1956, box 5, NSF/ODSF.

81. House, *Hearings before the Subcommittee on Independent Offices,* 30 January 1956, 528, 522.

82. Harry C. Kelly, "NSF Activities in Scientific and Technical Personnel and Education," report presented to the NCDSE, 18 September 1956, box 26, Waterman Papers.

83. Kelly interview, PSSC/OHC.

84. House, *Hearings before the Subcommittee on Independent Offices,* 9 February 1955, 235; 30 January 1956, 514.

85. "Education Held Threat to Soviets," *NYT,* 2 June 1955.

86. Presentation before the NSC, "Maintenance of Technological Superiority," 31 May 1956, box 7, AW/NSC.

87. Flemming to Eisenhower, memo outlining statement on formation of NCDSE, 29 February 1956, box 1, PCSE. The actual statement the president made announcing the formation of the National Committee was toned down considerably, see Eisenhower, *Public Papers,* vol. 4, 365–366.

88. Neil Carrothers to Waterman, 2 March 1956, box 18, NSF/ODSF.

89. Kelly, "NSF Activities in Scientific and Technical Personnel and Education," 18 September 1956, box 26, Waterman Papers.

90. Kelly to Waterman, Report on "The Development of Scientists and Engineers," 15 December 1956, box 18, NSF/ODSF.

91. SPE Divisional Committee meeting minutes, 10 November 1954, NSF/HF.

92. Kelly diary note, 8 March 1956, box 18, NSF/ODSF.

93. Robert L. Clark to Waterman, 10 May 1956, box 4, PCSE.

94. See, for example, "First Meeting of the National Committee for the Development of Scientists and Engineers," 15 May 1956, box 4; and NCDSE,

"Report of the Working Group on Long-Range Improvement of Science and Mathematics Programs in Elementary and Secondary Schools," June 1956, box 1, PCSE; also see Howard L. Bevis, "The National Committee for the Development of Scientists and Engineers," in *Scientific Manpower Report—1956* (Washington, D.C.: National Science Foundation, 1956), 11–15.

95. Allen W. Dulles to Flemming, 24 May 1956, box 12, PCSE.

96. SPE Divisional Committee meeting minutes, 10 November 1954, NSF/HF.

97. Kelly diary note, 8 March 1956, box 18, NSF/ODSF.

98. Kelly, "NSF Activities in Scientific and Technical Personnel and Education," 18 September 1956, box 26, Waterman Papers.

99. NSF internal memo, 20 June 1956, box 21, PCSE.

100. SPE Divisional Committee meeting minutes, 6 September 1956, NSF/HF; Bevis, "The National Committee for the Development of Scientists and Engineers," 11–15; NCDSE meeting minutes, 18 September 1956, NAS/ABE.

101. NSF, *Sixth Annual Report* (Washington, D.C.: U.S. Government Printing Office, 1956), 67–68; NSF, *Seventh Annual Report* (Washington, D.C.: U.S. Government Printing Office, 1957), 71.

102. Kelly interview, PSSC/OHC.

103. SPE Divisional Committee meeting minutes, 9 February 1956, NSF/HF.

## Chapter 4

1. Zacharias to Killian, 15 March 1956, box 170, C/K Papers; Kelly diary note, 14 July 1956, box 19, NSF/ODSF; Zacharias interview, PSSC/OHC; James R. Killian, "A Bold Strategy to Beat Shortage," *Life,* 7 May 1956, 147–149.

2. Kelly diary note, 14 July 1956, box 19, NSF/ODSF.

3. Zacharias interview, PSSC/OHC.

4. This chapter incorporates material from a previously published article entitled, "From World War to Woods Hole: The Use of Wartime Research Models for Curriculum Reform," first published in *Teachers College Record.*

5. Bruce Hevly, "Reflections on Big Science and Big History," in *Big Science: The Growth of Large-Scale Research,* ed. Peter Galison and Bruce Hevly (Stanford: Stanford University Press, 1992), 355; James H. Capshew and Karen A. Rader, "Big Science: Price to the Present," *Osiris,* 2d series, 7 (1992): 3–25.

6. A. Hunter Dupree, "The Great Instauration of 1940: The Organization of Scientific Research for War," in *The Twentieth-Century Sciences: Studies in the Biography of Ideas,* ed. Gerald Holton (New York: W.W. Norton & Company, 1972), 457–459.

7. Daniel J. Kevles, *The Physicists: The History of a Scientific Community in Modern America* (Cambridge: Harvard University Press, 1987), 302–308; Jack S.

Goldstein, *A Different Sort of Time: The Life of Jerrold Zacharias* (Cambridge: MIT Press, 1992), 50–59; and Charles Thorpe and Steven Shapin, "Who Was J. Robert Oppenheimer: Charisma and Complex Organization," *Social Studies of Science* 30 (2000): 545–548.

8. Goldstein, *Different Sort of Time,* 50–59; Peter Galison, "Physics between War and Peace," in *Science, Technology, and the Military,* ed. Everett Mendelsohn, Merit Roe Smith, and Peter Weingart (Dordrecht: Kluwer, 1988), 47–86, describes this approach as characteristic of much of the research undertaken during the war.

9. It is important to note that these newer scientific approaches were quite distinct from Taylorism and its allies. Scholars making this distinction include: Thomas Hughes, *Rescuing Prometheus* (New York: Pantheon, 1998), 8–9, 302; Michael Fortun and Sylvan S. Schweber, "Scientists and the State: The Legacy of World War II," in *Trends in the Historiography of Science,* ed. Kostas Gavroglu (Dordrecht: Kluwer, 1994), 336–338; and Andy Pickering, "Cyborg History and the World War II Regime," *Perspectives on Science* 3 (1995): n22, 18. Perhaps most important for my argument here is the fact that the experts who articulated the newer approaches to postwar problems explicitly drew on the physicists' analytical models rather than the earlier models of scientific management. Indeed, it was thought that the systems approach could only be fully developed by those with training in the research methods of the physical sciences; see Committee on Operations Research, *Operations Research with Special Reference to Non-Military Applications* (Washington, D.C.: National Research Council, 1951), 3–4.

10. Goldstein, *Different Sort of Time,* 52; Fortun and Schweber, "Scientists and the State," 330; Kevles, *The Physicists,* 320.

11. Galison, "Physics between War and Peace," 73. For an extended discussion of this point, see Galison, *Image and Logic: A Material Culture of Microphysics* (Chicago: University of Chicago Press, 1997), 239–311.

12. Paul E. Klopsteg, "Support of Basic Research from Government," in *Symposium on Basic Research,* ed. Dael Wolfle (Washington, D.C.: AAAS, 1959), 193.

13. Goldstein, *Different Sort of Time,* 98–108.

14. J. R. Marvin and F. J. Weyl, "The Summer Study," *Naval Research Reviews* 19, no. 8 (1966): 2.

15. Goldstein, *Different Sort of Time,* 152.

16. Allan A. Needell, "Project Troy and the Cold War Annexation of the Social Sciences," in *Universities and Empire: Money and Politics in the Social Sciences during the Cold War,* ed. Christopher Simpson (New York: New Press, 1998) 3–38. For a discussion of the "Campaign of Truth," see Walter L. Hixson, *Parting the Curtain: Propaganda, Culture, and the Cold War, 1945–1961* (New York: St. Martin's Press, 1997), 1–27. Connections between academic psychologists, communication experts, and the government's psychological warfare efforts after the war are well documented in

Christopher Simpson, *Science of Coercion: Communication Research and Psychological Warfare, 1945–1960* (New York: Oxford University Press, 1994).

17. MIT, "Project Troy Report to the Secretary of State," foreword, vii, Troy Report Files.

18. MIT, "Project Troy Report to the Secretary of State," foreword, vii, Troy Report Files.

19. Needell, "Project Troy and the . . . Social Sciences," 13.

20. L.V. Berkner to Secretary of State, 27 December 1950, Troy Report Files.

21. Marvin and Weyl, "The Summer Study," 6; Goldstein, *Different Sort of Time,* 155–156.

22. W. Henry Lambright, *Powering Apollo: James E. Webb of NASA* (Baltimore: Johns Hopkins University Press, 1995), 80–81.

23. MIT news release, 10 December 1956, box 170, C/K Papers; Kelly diary note, 24 September 1956, box 18, NSF/ODSF; PSSC meeting minutes, 8 September 1956, box 1, PSSC/OHC.

24. Zacharias interview, PSSC/OHC.

25. See J. Merton England, *A Patron for Pure Science: The National Science Foundation's Formative Years, 1945–57* (Washington, D.C.: National Science Foundation, 1982), 227–254.

26. Zacharias interview, PSSC/OHC.

27. SPE Divisional Committee meeting minutes, 16 December 1956, NSF/HF; Elbert P. Little, memo, 21 December 1956, box 170, PSSC/OHC.

28. Stratton to Zacharias, 20 April 1956, box 170, C/K Papers.

29. Stratton to Kelly, 15 August 1956, box 17, Zacharias Papers.

30. NAS Conference on PSSC, meeting transcript, 22 April 1958, NAS/ABE.

31. James R. Killian, *The Education of a College President* (Cambridge: MIT Press, 1985), 166.

32. House Committee on Science and Technology, Subcommittee on Science, Research, and Technology, *The National Science Foundation and Pre-College Science Education: 1950–1975,* report prepared by Science Policy Research Division, 94th Cong., 2d sess., 1976, Committee Print, appendix C, 230.

33. Jerrold R. Zacharias, "Team Approach to Education," *AJP* 29 (1961): 348.

34. Paul E. Marsh, "The Physical Science Study Committee: A Case History of Nationwide Curriculum Development, 1956–1961" (Ph.D. diss., Harvard University, 1964), 45–47.

35. Zacharias to various foundations, 18 June 1957, box 17, Zacharias Papers.

36. House Committee on Appropriations, *Hearings before the Subcommittee on Independent Offices,* 84th Cong., 1st sess., 9 February 1955, 234–235, emphasis added.

37. For membership of NSF hierarchy, see NSF, *Sixth Annual Report* (Washington, D.C.: U.S. Government Printing Office, 1956); James R. Killian, *Sputnik, Scientists, and Eisenhower: A Memoir of the First Special Assistant to the President for Science and Technology* (Cambridge: MIT Press, 1977), 277–279; Zacharias interview, PSSC/OHC.

38. Kelly interview, PSSC/OHC.

39. C. E. Sunderlin to Waterman, 20 November 1956, box 19, NSF/ODSF; Kelly to Waterman, "Principles for Support of Curriculum Development," memo, 26 June 1958, box 39, NSF/ODSF; Kevin Smith interview, PSSC/OHC.

40. PSSC meeting minutes, 8 September 1956, box 1, PSSC/OHC.

41. PSSC Report, 12 March 1957, box 16, Zacharias Papers; PSSC December conference minutes, 10 December 1956, box 19, NSF/ODSF.

42. U. Liddel to Kelly, 27 February 1957, box 33, NSF/ODSF.

43. "Progress Report: The First Six Years of PSSC, the First Four Years of ESI," 1966 (?), unpublished typescript, box 9, Zacharias Papers; Kelly diary note, 8 July 1957, box 33, NSF/ODSF.

44. Zacharias to D. R. Corson, 26 December 1957, box 33, NSF/ODSF; Zacharias interview, PSSC/OHC; Paul Boyer, *By the Bomb's Early Light: American Thought and Culture at the Dawn of the Atomic Age* (Chapel Hill: University of North Carolina Press, 1994), 77–78.

45. Zacharias interview, PSSC/OHC.

46. "Progress Report: The First Six Years of PSSC," box 9, Zacharias Papers.

47. E. C. Wine to J. E. Webb, 8 May 1959, Webb Papers.

48. David Hounshell, "The Cold War, RAND, and the Generation of Knowledge, 1946–1962," *HSPS* 27 (1997): 237–267; Fortun and Schweber, "Scientists and the State," 327–338; Hughes, *Rescuing Prometheus,* 15–67; Comm. on Operations Research, *Operations Research . . . Non-Military Applications,* 4–10.

49. S. S. Schweber, "The Mutual Embrace of Science and the Military: ONR and the Growth of Physics in the United States after World War II," in *Science, Technology, and the Military,* ed. Mendelsohn, et al., 28–32; Hughes, *Rescuing Prometheus,* 25–26.

50. Project Troy Report, viii-ix, 3, Annex 11, Troy Report Files.

51. Zacharias interview, PSSC/OHC.

52. SAC-ODM meeting summary, 12 May 1951; F. B. Llewellyn, "Systems Engineering with Reference to Its Military Applications," unpublished typescript, 1 November 1951; and SAC-ODM meeting summary, 8 February, 1952, box 194, C/K Papers.

53. Killian to various foundations, 11 September 1957, box 17, Zacharias Papers.

54. SPE Divisional Committee meeting minutes, 16 December 1956, NSF/HF.

55. Pickering, "Cyborg History," 21–23.

56. C. R. Carpenter, "A Challenge for Research," *Educational Screen* 27 (1948): 119–121; Arthur A. Lumsdaine, "Mass Communication Experiments in Wartime and Thereafter," *Social Psychology Quarterly* 47 (1984): 198–206; Ellen Herman, *The Romance of American Psychology: Political Culture in the Age of Experts* (Berkeley: University of California Press, 1995), 69–72.

57. Programmed instruction, which emerged from the work of B. F. Skinner, was another technology that many felt held some promise for improving education.

58. Henry Chauncey, "Teaching Personnel: More Effective Utilization of Teachers," in National Manpower Council, *Proceedings of a Conference on The Utilization of Scientific and Professional Manpower, October 7–11, 1953* (New York: Columbia University Press, 1954), 139–146.

59. Stanley Salmen to Zacharias, 10 August 1956, 13 August 1956, box 17, Zacharias Papers; PSSC meeting minutes, 10 November 1956, box 1, PSSC/OHC; Goldstein, *Different Sort of Time,* 173.

60. Herman, *Romance of American Psychology,* 69–71.

61. PSSC meeting minutes, 8 September 1956, box 1, PSSC/OHC; PSSC meeting minutes, 20 October 1956, box 170, C/K Papers.

62. Elbert Little to Waterman, 20 December 1956, box 19, NSF/ODSF; Zacharias to Tuttle, 30 January 1957, box 3, Jerome Wiesner Papers, MIT Archives.

63. Zacharias to Webb, 14 November 1958, box 362, Webb Papers.

64. Bruner to Apparatus of Teaching Group, 12 September 1959, NAS/ABE.

65. Zacharias to Killian, 15 March 1956, box 170, C/K Papers.

66. Physical Science Study Committee, *PSSC Physics: Teacher's Resource Book and Guide* (Boston: D.C. Heath, 1960).

67. Randall Whaley to Bronk, 2 July 1959, NAS/ABE. The NAS provided both the facilities at the Whitney Estate and its expertise to the Air Research and Development Command over the previous two summers for the study of potential Air Force weapon systems; Michael H. Gorn, *Harnessing the Genie: Science and Technology Forecasting for the Air Force, 1944–1986* (Washington, D.C.: Office of Air Force History, U.S. Air Force, 1988), 63–78.

68. Whaley, "Proposal for Summer Study on Fundamental Processes in Education," 29 April 1959, NAS/ABE.

69. Whaley to Bronk, 22 January 1959, NAS/ABE.

70. Bruner, "Notes for the Woods Hole Conference," 22 July 1959, NAS/ABE.

71. Jerome Bruner, *In Search of Mind: Essays in Autobiography* (New York: Harper and Row, 1983), 45; Burnham Kelly to Killian, 20 October 1950, box 4, Records of the Office of the Chancellor (Stratton), MIT Archives; Woods Hole study group participants, 9 September 1959, box 41, Carpenter Papers.

72. Whaley, "Proposal for Summer Study on Fundamental Processes in Education," 29 April 1959, NAS/ABE.

73. Summer Study on Educational Processes, tentative schedule, 9 [?] September, 1959; Bruner, remaining agenda, 14 September 1959, NAS/ABE. The panel members included psychologists C. Ray Carpenter, John B. Carroll, and Donald Taylor; John A. Fischer from Teachers College; and film experts John Flory and Don Williams.

74. Panel Report/Apparatus of Teaching, 18 September 1959, NAS/ABE.

75. Panel Report/Apparatus of Teaching, 18 September 1959, NAS/ABE.

76. Bruner to Apparatus of Teaching Group, 12 September 1959, NAS/ABE.

77. Arthur W. Melton, Report of the NAS-NRC Symposium on Education and Training Media, 18–19 August 1959, Glass Papers.

78. Philip Morrison, "Insight and Taste," *AJP* 31 (1963): 477.

79. Zacharias interview, PSSC/OHC.

80. AIBS Executive Committee meeting minutes, 4 January 1952, box 2, AIBS papers. NRC, Conference on Biological Education meeting minutes, 10 March 1953, NAS/CEP.

81. CEP meeting minutes, 13 November 1954, NAS/CEP; see also Richard E. Paulson, "The Committee on Educational Policies," *AIBS Bulletin* 5, no. 5 (1955): 17–18.

82. Frank L. Campbell, "Committee on Educational Policies in Biology," 1958, NAS/CEP; Chester A. Lawson and Richard E. Paulson, *Laboratory and Field Studies in Biology: A Sourcebook for Secondary Schools* (New York: Holt, Rinehart and Winston, 1958).

83. Richard Paulson, memo re: conversation with J. Behnke, 12 February 1958; Paulson to Behnke, 5 December 1957; Paulson to H. M. Phillips, 18 January 1956, NAS/CEP.

84. CEP recommendation, 15 March 1958, NAS/CEP.

85. Kelly diary note, 13 November 1958, box 2, NSF Assoc. Dir. Papers.

86. SPE Divisional Committee meeting minutes, 6 May 1958, NSF/HF.

87. Zacharias interview, PSSC/OHC.

88. Paulson to Behnke, 5 December 1957, emphasis added, NAS/CEP.

89. Kelly to Waterman, "Principles for Support of Curriculum Development," memo, 26 June 1958, box 39, NSF/ODSF.

90. Kelly to Loeb, 7 March 1957, NSF/HF.

91. Paulson, memo re: conversations with NABT, 12 February 1958, NAS/CEP.

92. Zacharias to Kelly, 16 April 1958, box 39, NSF/ODSF.

93. SPE Divisional Committee meeting minutes, 6 May 1958, NSF/HF; Kelly diary note, 23 September 1958, box 33, NSF/ODSF. The formal decision to submit a proposal to NSF came at the meeting of the AIBS Governing Board on May 9, 1958, Governing Board meeting minutes, 9 May 1958, box 9, AIBS papers.

94. Richard Paulson diary note, 6 October 1958, box 2, NSF Assoc. Dir. Papers.

95. Paulson, "Proposals and Grants for the Biological Sciences Curriculum Study," NSF internal document, 1963 (?), box 69, NSF/ODSF; Kelly diary note, 16 December 1958; Litwin diary note, 28 January 1959, box 2, NSF Assoc. Dir. Papers. The five biologists from the "Zach" list were: H. Burr Steinbach, David R. Goddard, Paul B. Sears, Wallace O. Fenn, and Alfred S. Romer.

96. BSCS Steering Committee meeting minutes, 5 February 1959, BSCS Papers.

97. Richard E. Paulson to Glass, 22 September 1960, Glass Papers.

98. D. Goddard to Glass, 21 March 1960, BSCS Papers.

99. Glass to A. Lee, 27 February 1960, BSCS Papers.

100. Zacharias to Killian, 15 November 1960, box 31, Killian Papers.

101. Webb to Killian, 27 July 1960, box 31, Killian Papers.

102. Kelly to Frederic H. Bohnenblust, 22 October 1958, box 39, NSF/ODSF.

103. For scandal, see J. Merton England's, "The National Science Foundation and Curriculum Reform: A Problem of Stewardship," *Public Historian* 11 (1989): 23–36; P. Handler to Waterman, 21 January 1963, NSF/HF; for missteps, see McCulloch to Hartgering, 2 April 1963, box 255, OST.

104. Paulson, "Proposals and Grants for the Biological Sciences Curriculum Study," NSF internal document, 1963 (?), box 69, NSF/ODSF; Kelly to Glass, 30 June 1960, BSCS Papers.

105. Robert A. Divine, *Blowing on the Wind: The Nuclear Test Ban Debate, 1954–1960* (New York: Oxford University Press, 1978), 135; see also Bentley Glass, "The Academic Scientist, 1940–1960," *Science* 132 (1960): 598–603; National Research Council, *The Biological Effects of Atomic Radiation: A Report from a Study* (Washington, D.C.: NAS-NRC, 1956).

106. Cox to Killian, 31 January 1958, box 5, OSAST.

107. Webb to Killian, 27 July 1960, box 31, Killian Papers.

108. BSCS Steering Committee meeting minutes, 5 February 1959, BSCS Papers.

109. Kelly to Waterman, 26 June 1958, box 39, NSF/ODSF; Kelly to Webb, 11 May 1959, box 19, Zacharias Papers; quotation from Kelly diary note, 16 December 1958, box 2, NSF Assoc. Dir. Papers.

110. Glass to Waterman, 22 January 1963, box 69, NSF/ODSF; quotation from Cox to Glass, 26 April 1960, BSCS Papers.

111. Grobman, "The BSCS," unpublished report, spring 1959 (?), box 9, Bates Papers.

112. AIBS, funding proposal, 29 April 1960, BSCS Papers.

113. Cox to Bates, 17 September 1958, box 10, Bates Papers; BSCS statement on film program, 7 May 1962, BSCS Papers.

114. BSCS Steering Committee meeting minutes, 28 January 1960, BSCS Papers.

115. BSCS statement on film program, 7 May 1962, BSCS Papers.

116. AIBS, funding proposal, 29 April 1960, BSCS Papers.

117. BSCS Course Content Committee meeting report, 30 October 1959; BSCS Steering Committee meeting minutes, 28 January 1960, BSCS Papers. Moore to Bates, 10 February 1960, box 9, Bates Papers.

118. Moore to Bates, 10 February 1960, box 9, Bates Papers.

119. Grobman to W. Fenn, 25 February 1960, BSCS Papers.

120. BSCS Steering Committee meeting minutes, 26 May 1962, BSCS Papers.

121. Grobman to BSCS Executive Committee, 14 March 1960, BSCS Papers.

122. SPE Divisional Committee meeting minutes, 15 September 1960, NSF/HF.

123. Arnold B. Grobman, *The Changing Classroom: The Role of the Biological Sciences Curriculum Study* (Garden City, NY: Doubleday & Company, 1969), 2.

124. Diane Ravitch, *The Troubled Crusade: American Education, 1945–1980* (New York: Basic Books, 1983), 229; Philip W. Jackson, "The Reform of Science

Education: A Cautionary Tale," *Daedalus* 112 (1983): 147; and Frank G. Jennings, "It Didn't Start with *Sputnik,*" *Saturday Review,* 16 September 1967, 96.

125. In addition to aiding the scientists' reform efforts in education, *Sputnik* catalyzed federal support of NSF's longstanding push for increased funds and support for basic research. Roger L. Geiger, "What Happened After *Sputnik?* Shaping University Research in the United States," *Minerva* 35 (1997): 349–367.

126. For a complete historical account of the political events precipitated by *Sputnik,* and President Eisenhower's management of the crisis, see Robert A. Divine, *The Sputnik Challenge: Eisenhower's Response to the Soviet Satellite* (New York: Oxford University Press, 1993); and Killian, *Sputnik, Scientists, and Eisenhower.*

127. Zacharias to D. McGee, 22 October 1957, box 16, Zacharias Papers.

128. Michael Amrine, "The Meaning of the Red Moons," *Progressive,* December 1957, 13.

129. "Crisis in Education," *Life,* 24 March 1958; the other parts of the series were published on March 31, April 7, April 14, and April 21, 1958; "What Price Life Adjustment?" *Time,* 2 December 1957, 53; Sloan Wilson, "It's Time to Close Our Carnival," *Life,* 24 March 1958, 36; "The Deeper Problem in Education," *Life,* 31 March 1958, 32.

130. "New Cabinet Post Eisenhower 'Ideal'," *NYT,* 11 March 1959.

131. "The 3 R's in Russia Are Really Tough," *USN,* 15 November 1957, 137; "The Fighting Challenge of Russia's Schools," *Look,* 14 October 1958, 38–42; John Gunther, "Russia Rings the School Bell," *Readers' Digest,* March 1959, 176–179; Hyman G. Rickover, *Education and Freedom* (New York: Dutton, 1959). Interestingly enough, during the height of the *Sputnik* crisis a Gallup poll showed that, although 90 percent of school administrators stated schools demanded too little work from students, only 51 percent of parents shared that sentiment, suggesting that pressure for reform came from the top rather than from the grassroots.

132. Dwight D. Eisenhower, *Public Papers of the Presidents of the United States: Dwight D. Eisenhower, 1953–1961,* vol. 5 (Washington, D.C.: U.S. Government Printing Office, 1958), 796.

133. W. A. Jaracz diary note, 16 January 1959, box 2, NSF Assoc. Dir. Papers; see also Dwight D. Eisenhower, "Recommendations Relative to Our Educational System," *Science Education* 42 (1958):103–106.

134. Goodpaster memcon, 15 October 1957 meeting with SAC, box 23; PSAC to Goodpaster, 6 November 1957, box 23, WHSS.

135. For a discussion of membership and origins of PSAC, see Killian, *Sputnik, Scientists, and Eisenhower.* For science advising more generally, see Gregg Herken, *Cardinal Choices: Presidential Science Advising from the Atomic Bomb to SDI* (New York: Oxford University Press, 1992).

136. Eisenhower, *Public Papers,* vol. 5, 776; "President to Aid Drive for Science," *NYT,* 31 October 1957.

137. Waterman to Eisenhower, 20 November 1957, box 4, PSAC.

138. Divine, *Sputnik Challenge,* 81, 157.

139. Internal memo on NSF relationship with U.S. Office of Education, 28 March 1956, NSF/HF; NSF meeting summary, NSF-OE, 24 July 1957, box 25, NSF/ODSF. The U.S. Office of Education prior to *Sputnik* had very little involvement in curriculum development, its primary task historically being the collection of educational statistics. What little curriculum work there was was targeted to specific immediate need-type areas, such as civil defense education, see, for example, JoAnne Brown, "'A is for Atom, B is for Bomb': Civil Defense in American Public Education, 1948–1963," *Journal of American History* 75 (1988): 68–90.

140. SPE Divisional Committee meeting minutes, 14 November 1957, box 52, NSF/ODSF.

141. Divine, *Sputnik Challenge,* 89–90; J. Stambaugh to Killian, 26 November 1957, box 8, OSAST.

142. For a complete legislative history of the National Defense Education Act, see Barbara Barksdale Clowse, *Brainpower for the Cold War: The Sputnik Crisis and National Defense Education Act of 1958* (Westport, Conn.: Greenwood Press, 1981). The full text of the NDEA can be found in the appendix of Clowse's book.

143. "Do-It-Yourself Physics," *Life,* 14 April 1958, 122–123; Bess Furman, "School Aid Bills to Stress Science," *NYT,* 26 December 1957; Thomas M. Pryor, "School Science Gets a New Look," *NYT,* 17 November 1957; "The New Physics Class," *Time,* 17 March 1958, 41–42.

144. Other high school science curriculum projects included the chemistry programs, Chemical Bond Approach (CBA) and Chemical Education Materials Study (CHEM Study). In mathematics the most well-known program was the intermediate-level School Mathematics Study Group (SMSG). The elementary school social studies project, Man: A Course of Study (MACOS), was, perhaps, the most controversial of all the new curricula. First-person histories of SMSG, CHEM Study, and MACOS are told in William Wooton, *SMSG: The Making of a Curriculum* (New Haven: Yale University Press, 1965); Richard J. Merrill and David W. Ridgway, *The CHEM Study Story* (San Francisco: W.H. Freeman and Company, 1969); and Peter B. Dow, *Schoolhouse Politics: Lessons from the Sputnik Era* (Cambridge: Harvard University Press, 1991). A complete list of NSF-funded curriculum projects can be found in NSF, *Course and Curriculum Improvement Materials* (Washington, D.C.: U.S. Government Printing Office, 1976).

145. L. DuBridge to Flemming, 9 September 1958, box 8, OSAST.

146. J. Boyer Jarvis, "The Role of the United States Office of Education in Curriculum Experimentation," speech, 27 September 1961, box 3, USOE Sel. Central Files. See also, J. J. McPherson, "Let's Look at the Systems Concept of Educational Planning," *High School Journal* 44 (1960): 66–72.

147. PSAC Education Panel meeting summary, 1 March 1962, box 138, OST.

148. Purcell interview, PSSC/OHC.

## Chapter 5

1. SPE Divisional Committee meeting minutes, 9 February 1956, NSF/HF.

2. National Science Board, executive session meeting minutes, 18 June 1957, NSF/HF.

3. Alan T. Waterman, "Science in American Life and in the Schools," in *The High School in a New Era,* ed. Francis S. Chase and Harold A. Anderson (Chicago: University of Chicago Press, 1958), 84–85.

4. George B. Kistiakowsky to James E. Ruddell, 29 August 1960, box 8, OSAST.

5. The contents of the New York state physics syllabus are described in American Institute of Physics, *Physics in Your High School* (New York: McGraw-Hill, 1960), 56–59; PSSC memo, "What We Seem to Know," 11 October 1956, box 1, PSSC/OHC; PSSC memo, "What We Seem to Know," 6 November 1956, box 170, C/K Papers.

6. PSSC memo, "What We Seem to Know," 11 October 1956, box 1, PSSC/OHC.

7. Walter C. Michels, "High School Physics: A Report of the Joint Committee on High School Teaching Materials," *Physics Today* 10 (1957): 20–21; Michels report to PSSC, "Committee on High School Teaching Materials: Comments by Walter Michels," 31 May 1956, box 1, PSSC/OHC. The impetus behind this examination of high school teaching materials came from the Greenbriar Conference on the Production of Physicists convened in the spring of 1955 and sponsored by the Division of Physical Sciences of the National Research Council, see "NRC, AIP Conference on the Production of Physicists," *Physics Today* 8 (1955): 6–12; Michels to AAPT Executive Committee, "Recommendations and Requests of the Greenbriar Conference," 14 April 1955, NAS/CPP.

8. Michels report to PSSC, "Committee on High School Teaching Materials," 31 May 1956, box 1, PSSC/OHC; PSSC memo, "What We Seem to Know," 6 November 1956, box 170, C/K Papers.

9. These examples are taken from Charles E. Dull, H. Clark Metcalfe, and William O. Brooks, *Modern Physics* (New York: Henry Holt and Company, 1955), 247, 310, 531. Similar examples are pervasive in the popular physics textbooks of the time.

10. PSSC Report, 12 March 1957, box 16, Zacharias Papers.

11. Michels report to PSSC, "Committee on High School Teaching Materials," 31 May 1956, box 1, PSSC/OHC.

12. Quotation from PSSC, "First Annual Report of the Physical Science Study Committee: Preliminary Edition," ca. 27 January 1958, p. 4, box 6, PSSC Papers; PSSC December conference minutes, 10 December 1956, box 19,

NSF/ODSF; PSSC memo, "What We Seem to Know," 6 November 1956, box 170, C/K Papers.

13. PSSC Report, 12 March 1957, box 16, Zacharias Papers.

14. Zacharias interview, PSSC/OHC.

15. PSSC December conference minutes, 10 December 1956, box 19, NSF/ODSF.

16. NSF proposal, 17 August 1956; also Stanley Salmen to Zacharias, 10 August 1956, box 17, Zacharias Papers.

17. PSSC, "First Annual Report," p. 4, box 6, PSSC Papers.

18. SPE Divisional Committee meeting minutes, 16 December 1956, NSF/HF.

19. Urner Liddel to Kelly, memo, 27 February 1957, box 33, NSF/ODSF; Zacharias interview, PSSC/OHC. The more widely disseminated of the new chemistry curricula was CHEM Study, directed by Glenn T. Seaborg, Richard J. Merrill, and David W. Ridgway, *The CHEM Study Story* (San Francisco: W.H. Freeman and Company, 1969).

20. NSF, *Basic Research: A National Resource* (Washington, D.C.: U.S. Government Printing Office, 1957), vii.

21. NSF, *Basic Research,* 5.

22. Dael Wolfle, ed., *Symposium on Basic Research* (Washington, D.C.: American Association for the Advancement of Science, 1959).

23. PSSC, *Physics* (Boston: D.C. Heath, 1960), 4.

24. For previous curricular norms in science, see, for example, A. N. Zechiel, "Recent Trends in Revision of Science Curricula," *Educational Method* 16 (1937): 402–407; Hal Baird, "Teaching the Physical Sciences from a Functional Point of View," *Educational Method* 16 (1937): 407–412; Department of Science Instruction, *Science Instruction and America's Problems* (Washington, D.C.: National Education Association, 1940); and "Science in Secondary Schools Today," a special issue of the *Bulletin of the National Association of Secondary-School Principals* 37, no. 191 (1953).

25. Daniel Lee Kleinman and Mark Solovey, "Hot Science/Cold War: The National Science Foundation after World War Two," *Radical History Review* 63 (1995): 110–139.

26. Jerrold R. Zacharias, "Educational Methods and Today's Science—Tomorrow's Promise," *Technology Review* 59 (1957): 501.

27. Graham DuShane, "The Spirit of Science," *Science* 123 (1956): 165; Howard Mumford Jones, "A Humanist Looks at Science," in *Science and the Modern Mind,* ed. Gerald Holton (Boston: Beacon Press, 1958), 100–108; Charles Morris, "Prospects for a New Synthesis: Science and the Humanities as Complimentary Activities," in *Science and the Modern Mind,* ed. Holton, 92–99; J. Robert Oppenheimer, "Science and Modern Society," *New Republic,* 6 August 1956, 8–9; I. I. Rabi, "Scientist and Humanist: Can the Minds Meet?" *Atlantic Monthly,* January 1956, 64–67; Eugene Rabinowitch, "Science and Humanities in Education," *Bulletin of the American Association of University Professors* 44 (1958): 449–464.

28. Elbert P. Little, "The Physical Science Study: From these Beginnings," *Science Teacher* 24, no. 7 (1957): 316.

29. J. A. Stratton, "Science and the Educated Man," *Physics Today* 9 (1956): 19–20.

30. Charles Frankel quoted in Little, "From these Beginnings," 318.

31. PSSC, "General Report of the Physical Science Study Committee," 20 August 1957, box 17, Zacharias Papers.

32. Judson Cross, "Notes on Teaching the New Physics Course," 8 March 1958, box 1, PSSC/OHC.

33. Bates to Harry Fuller, 7 May 1954, box 6, Bates Papers.

34. Paulson to Richard T. Wareham, 24 October 1955, NAS/CEP.

35. Explicit links to *How We Think* can be found in teacher-training books published throughout the first half of the century, see, for example, Elliot Rowland Downing, *Teaching Science in the Schools* (Chicago: University of Chicago Press, 1925), 53–56; Elwood D. Heiss, Ellsworth S. Obourn, and Charles W. Hoffman, *Modern Science Teaching* (New York: Macmillan, 1950), 93; and M. Louise Nichols, "The High School Student and Scientific Method," *Journal of Educational Psychology* 20 (1929): 196–204. James B. Conant attributed the source of the "alleged scientific method" to Karl Pearson's book, *The Grammar of Science* (1892), which presents a similar stepwise approach to science. Few educators, though, seem to have drawn on that particular source.

36. John Dewey, *How We Think* (Boston: D.C. Heath, 1910), 72.

37. Alan Ryan, *John Dewey and the High Tide of American Liberalism* (New York: W. W. Norton, 1995), 72–73, 142–146; John Dewey, "Philosophy's Future in Our Scientific Age," *Commentary* 8 (1949): 388–394; see also Robert B. Westbrook, *John Dewey and American Democracy* (Ithaca: Cornell University Press, 1991), 141–143.

38. National Society for the Study of Education (NSSE), *Science Education in American Schools: Forty-Sixth Yearbook of the NSSE* (Chicago: University of Chicago Press, 1947), 20.

39. The Harvard Report, *General Education in a Free Society,* included this comment on the scientific method: "[N]othing is further from the procedure of the scientist than a rigorous tabular progression through the supposed 'steps' of the scientific method, with perhaps the further requirement that the student not only memorize but follow this sequence in his attempt to understand natural phenomena," Committee on the Objectives of General Education in a Free Society, *General Education in a Free Society* (Cambridge: Harvard University Press, 1945), 158. Zacharias referred to these pages as providing a useful starting point for understanding the nature of science; Zacharias to Killian, 15 March 1957, box 170, C/K Papers.

40. Conant's goal was just this, getting the interested lay public to understand the process of science as a set of principles regarding the tactics and strategy of science. He thought the military terminology was particularly descriptive of the methods scientists used and would be well understood by

the public following World War II; James B. Conant, *On Understanding Science* (New Haven: Yale University Press, 1949), 98–109.

41. See Paul DeH. Hurd, "The Scientific Method as Applied to Personal-Social Problems," *Science Education* 39 (1955): 262–265.

42. Percy W. Bridgman, *Reflections of a Physicist* (New York: Philosophical Library, 1950), 351; Judson Cross, "Notes on Teaching the New Physics Course," 8 March 1958, box 1, PSSC/OHC. This view of the scientific method was shared by some of the biologists as well, see BSCS Steering Committee meeting minutes, 13 May 1961, BSCS Papers.

43. Zacharias interview, PSSC/OHC.

44. PSSC, "General Report of the Physical Science Study Committee," 20 August 1957, box 17, Zacharias Papers; Philip Morrison stated that although "to some extent we were pressed to make it a way to recruit scientists, we never—at least I never—accepted that as an important criterion." Morrison interview, 28 August 1975, PSSC/OHC.

45. Zacharias interview, PSSC/OHC.

46. Paulson to Richard T. Wareham, 24 October 1955, NAS/CEP.

47. The scientific community, in the spirit of the Arden House Conference, continued its efforts to improve the social standing of science in the United States during this time. The AAAS appointed a special committee to examine the social aspects of science, which made its report in December 1956, see Robert K. Plumb, "Scientists to Help Society Meet Impact of Advances," *NYT,* 31 December 1956; for the full text of the AAAS report, see "Social Aspects of Science: Preliminary Report of AAAS Interim Committee," *Science* 125 (1957): 143–147. The AAAS followed up this work with the convening of the Parliament of Science in March 1958, which had similar objectives, see AAAS, "1958 Parliament of Science," *Science* 127 (1958): 852–858.

48. William L. Laurence, "Science Depicted in Lowest Esteem," *NYT,* 2 February 1956.

49. Faculty meeting minutes, 21 January 1953, MIT Faculty Records, MIT Archives. On Zacharias conspiracy, see Atomic Energy Commission, *In the Matter of J. Robert Oppenheimer: Transcript of Hearing before Personnel Security Board and Texts of Principal Documents and Letters* (Cambridge: MIT Press, 1971), 600–601.

50. Wiesner quoted in Jack S. Goldstein, *A Different Sort of Time: The Life of Jerrold Zacharias* (Cambridge: MIT Press, 1992), 142.

51. Goldstein, *Different Sort of Time,* 141–142.

52. Paul Boyer, *By the Bomb's Early Light: American Thought and Culture at the Dawn of the Atomic Age* (Chapel Hill: University of North Carolina Press, 1994), 77–79; "How Communists Try to Influence American Teachers," *USN,* 31 July 1953, 83. A detailed account of Morrison's harassment can be found in Ellen W. Schrecker, *No Ivory Tower: McCarthyism and the Universities* (New York: Oxford University Press, 1986), 149–160.

53. "Pure Scientists Called Red Prey," *NYT,* 14 December 1956, emphasis added.

54. Morrison and Zacharias interviews, PSSC/OHC.

55. PSSC, "First Annual Report," p. 36, box 6, PSSC Papers.

56. PSSC meeting minutes, 20 October 1956, box 170, C/K Papers.

57. ATW diary note, 3 May 1957, box 25, NSF/ODSF.

58. Zacharias, "Educational Methods and Today's Science—Tomorrow's Promise," 503.

59. ATW diary note, 3 May 1957, box 25, NSF/ODSF.

60. PSSC Report, 12 March 1957, box 16, Zacharias Papers.

61. Goldstein, *Different Sort of Time,* 174–178.

62. Jerrold Zacharias, "A Scientist Looks at Scientists," in *The Scientific Revolution: Challenge and Promise,* ed. G. W. Elbers and P. Duncan (Washington, D.C.: Public Affairs Press, 1959), 134. The AAAS poll is described in Margaret Mead and Rhoda Métraux, "Image of the Scientist among High School Students: A Pilot Study," *Science* 126 (1957): 384–389.

63. PSSC, "First Annual Report," appendix 5, box 6, PSSC Papers.

64. PSSC meeting minutes, 20 October 1956, box 170, C/K Papers.

65. Zacharias interview, PSSC/OHC.

66. PSSC, *PSSC Physics: Teacher's Resource Book and Guide* (Boston: D.C. Heath, 1960), section 1–3.

67. PSSC, "First Annual Report," appendix 5, box 6, PSSC Papers.

68. Zacharias, "Educational Methods and Today's Science—Tomorrow's Promise," 503.

69. ATW diary note, 10 March 1959, box 25, NSF/ODSF.

70. Zacharias interview, PSSC/OHC.

71. *The Pressure of Light,* 16 mm (Newton, Mass.: Physical Science Study Committee, 1958).

72. SPE Divisional Committee meeting minutes, 14 November 1957, box 52, NSF/ODSF; Zacharias, "Educational Methods and Today's Science—Tomorrow's Promise," 503.

73. *Pressure of Light.*

74. Zacharias interview, PSSC/OHC.

75. Jerrold R. Zacharias, "Into the Laboratory," *Science Teacher* 24, no. 7 (1957): 324.

76. Zacharias interview, PSSC/OHC.

77. Judson Cross, "Notes on Teaching the New Physics Course," 8 March 1958, box 1, PSSC/OHC.

78. PSSC, *Physics Laboratory Guide No. 1, Preliminary Edition* (Cambridge: Physical Science Study Committee, 1958), 1–2.

79. PSSC, *Physics Laboratory Guide No. 1,* 1–2.

80. PSSC, *Laboratory Guide for Physics* (Boston: D.C. Heath, 1960), iv; PSSC, *Teacher's Guide for Laboratory Experiments* (Boston: D.C. Heath, 1960), 1.

81. PSSC, *Physics Laboratory Guide No. 1,* 1–2.

82. James B. Conant, ed., *Harvard Case Histories in Experimental Science,* vol. 1 (Cambridge: Harvard University Press, 1964), viii.

83. Peter Galison, *Image and Logic: A Material Culture of Microphysics* (Chicago: University of Chicago Press, 1997), 307.

84. PSSC, *Physics Laboratory Guide No. 1,* 1–1.

85. PSSC, *Physics Laboratory Guide No. 1,* 1–1.

86. PSSC, *Laboratory Guide for Physics,* iv.

87. PSSC, *Laboratory Guide for Physics,* iv.

88. Zacharias, "Into the Laboratory," 325.

89. Zacharias, "Into the Laboratory," 324.

90. The origins and influence of operationalism in the twentieth century, formally advanced by Harvard physicist Percy Bridgman, are described in Maila L. Walter, *Science and Cultural Crisis: An Intellectual Biography of Percy Williams Bridgman (1882–1961)* (Stanford: Stanford University Press, 1990).

91. Lab meeting minutes, 25 May 1959, box 3, PSSC Papers.

92. Jean Marean to Gilbert Finley, memo, 3 February 1958, box 17, Zacharias Papers.

93. Lab meeting minutes, 5 November 1959, box 3, PSSC Papers.

94. There was always a great deal of overlap among scientists on the various national committees, professional panels, and government advisory bodies. PSSC and PSAC were no different. PSSC-related representation on PSAC and PSAC's Education Panel included Zacharias, Stephen White, Elbert Little, Henry Chauncey, James Killian, Alan Waterman, Edwin Land, Edward Purcell, and I. I. Rabi.

95. PSAC Education Panel meeting minutes, 6 February 1958, 7 March 1958, 26 April 1958, box 14, OSAST; DuBridge to C. P. Haskins, Waterman, and Zacharias, 13 January 1958, box 188, DuBridge papers.

96. Zacharias, outline of role of the federal government in education, 2 January 1957, box 37, NSF/ODSF; PSSC, "General Report," 25 March 1957, box 17, PSSC Papers.

97. President's Science Advisory Committee, *Education for the Age of Science* (Washington, D.C.: U.S. Government Printing Office, 1959), 15.

98. Goodpaster memcon, 15 October 1957 meeting with SAC, box 23, WHSS; see also Henry Chauncey to Education Panel, 18 July 1958, box 14, OSAST.

99. AAAS, "Social Aspects of Science: Preliminary Report of AAAS Interim Committee," 144; on the economic boom of the late 1950s, see James T. Patterson, *Grand Expectations: The United States, 1945–1974* (New York: Oxford University Press, 1996), 313.

100. James R. Killian, "Maintaining the Technological Strength of the United States," speech before the Women's National Press Club, Washington, D.C., 8 January 1958, box 10, NDEA Files.

101. Killian, "The Challenge of Soviet Science," speech before the Manufacturing Chemists Assoc., New York City, 8 April 1957, box 195, C/K Papers.

102. Killian to DuBridge, memo, 23 May 1958, box 8; PSAC Education Panel meeting minutes, 7 March 1958, box 14, OSAST; PSAC, *Education for the Age of Science*, 7–8.

103. Morris Ernst to Maxwell Rabb, 19 November 1957, box 1191, WHCF.

104. PSSC, *Education for the Age of Science*, 18.

105. PSAC Education Panel meeting minutes, 21 July 1958, box 5, PSAC.

106. Walter C. Michels, "The Teaching of Elementary Physics," *Scientific American* 198 (1958): 57–58.

107. PSSC, *Education for the Age of Science*, 4, emphasis added.

108. PSSC, "General Report of the Physical Science Study Committee," 20 August 1957, box 17, Zacharias Papers.

109. AAAS, "Social Aspects of Science: Preliminary Report of AAAS Interim Committee," 143.

110. Zacharias interview, PSSC/OHC.

## Chapter 6

1. Robert K. Plumb, "Biologist Teams Rush Textbooks," *NYT,* 29 August 1960.

2. Bentley Glass, "Revolution in Biology," *BSCS Newsletter,* no. 9 (1961): 6–7.

3. Glass handwritten note, 2 March 1961, BSCS Papers; Zacharias to Killian, 15 November 1960, box 31, Killian Papers; McCulloch to Hartgering, 2 April 1963, box 255, OST; Oakley to Killian, 7 April 1961, box 31, Killian Papers.

4. Bentley Glass, "Liberal Education in a Scientific Age," *BAS* 14 (1958): 348–351. These ideas were developed more completely in his book, *Science and Liberal Education* (Baton Rouge: Louisiana State University Press, 1959).

5. Bentley Glass, "The Responsibilities of Biologists," *AIBS Bulletin* 7, no. 5 (1957): 13.

6. Glass, "Liberal Education in a Scientific Age," 349.

7. Glass, "Responsibilities of Biologists," 11.

8. Grobman, "The Biological Sciences Curriculum Study," unpublished report, ca. April 1959, box 9, Bates Papers.

9. Bentley Glass, "A Letter to All Biologists," *AIBS Bulletin* 4, no. 5 (1954): 5.

10. Alan T. Waterman, "Federal Support of Fundamental Research in the Biological Sciences," *AIBS Bulletin* 1, no. 5 (1951): 11–17. The preferential funding of the physical sciences over the biological sciences was also pointed to explicitly in the AAAS report on the place of science in society; AAAS, "Social Aspects of Science: Preliminary Report of AAAS Interim Committee," *Science* 125 (1957): 144.

11. Toby A. Appel, *Shaping Biology: The National Science Foundation and American Biological Research, 1945–1975* (Baltimore: Johns Hopkins University Press, 2000), 62. On fellowship funding, see NSF Annual Reports from 1950–1958.

12. Waterman to Lee Anna Embrey, memo, 25 October 1954, NSF/HF.
13. CEP meeting minutes, 21 May 1955, 13 November 1954, NAS/CEP.
14. Walter P. Taylor, "Is Biology Obsolete?" *AIBS Bulletin* 8, no. 3 (1958): 12–14.
15. Cox to Killian, 31 January 1958, box 5, OSAST.
16. Killian to Wanda, typed notes, 31 January 1958, box 5, OSAST.
17. Paul Weiss, "The Challenge of Biology," *AIBS Bulletin* 4, no. 2 (1954): 9.
18. Marston Bates, "Where is Biology?" *AIBS Bulletin* 2, no. 3 (1952): 15.
19. Marston Bates, "The Fragmentation of Biology," unpublished typescript, 1950, box 4, Bates Papers.
20. Grobman to Content Committee, 9 February 1960, box 9, Bates Papers; Truman J. Moon, Paul B. Mann, and James H. Otto, *Modern Biology*, rev. ed. (New York: Henry Holt and Company, 1951); Ella Thea Smith, *Exploring Biology*, 5th ed. (New York: Harcourt Brace, 1959). For a discussion of other textbooks BSCS surveyed, see Paul DeHart Hurd, *Biological Education in American Secondary Schools, 1890–1960,* BSCS Bulletin no. 1 (Washington, D.C.: American Institute of Biological Sciences, 1961), 195–204.
21. Grobman, "The Biological Sciences Curriculum Study," unpublished report, ca. April 1959, box 9, Bates Papers.
22. Herman T. Spieth to Glass, 14 March 1960, BSCS Papers.
23. Glass, "Responsibilities of Biologists," 10.
24. BSCS Steering Committee meeting minutes, 5 February 1959, BSCS Papers. Glass to *New York Times,* 22 June 1960, BSCS Papers. Glass's letter was published in edited form on 16 July 1960.
25. Grobman, "The Biological Sciences Curriculum Study," unpublished report, ca. April 1959, box 9, Bates Papers.
26. BSCS Steering Committee meeting minutes, 2 February 1961; BSCS Steering Committee meeting minutes, 9 June 1959, BSCS Papers.
27. BSCS, "History of the BSCS," *BSCS Newsletter,* no. 1 (1959): 4.
28. John A. Moore, "The Committee on the Content of the Curriculum," *BSCS Newsletter,* no. 2 (1960): 5.
29. Grobman, memo on summer writing conference, 8 April 1960, BSCS Papers.
30. Arnold B. Grobman, *The Changing Classroom: The Role of the Biological Sciences Curriculum Study* (Garden City, NY: Doubleday and Company, 1969), 81.
31. BSCS Committee on the Content of the Curriculum meeting report, 30 October 1959, BSCS Papers.
32. BSCS Steering Committee meeting minutes, 5 February 1959, 28 January 1960, BSCS Papers.
33. Grobman, *Changing Classroom,* 76.
34. The seven hierarchical levels of biological organization were: molecular, cellular, tissue and organ, individual organism, population, community, and world biome; Grobman, *Changing Classroom,* 73–82.
35. Glass to *New York Times,* 22 June 1960, BSCS Papers.
36. Philip J. Pauly, "The Development of High School Biology: New York City, 1900–1925," *Isis* 82 (1991): 662–668.

37. Oscar Riddle, "Amount and Nature of Biology Teaching in Secondary Schools," in *The Teaching of Biology in Secondary Schools of the United States,* ed. Oscar Riddle (Evanston: Committee on the Teaching of Biology of the Union of American Biological Societies, 1942), 64–66. Glass was one of six biologists who had conducted the survey.

38. Schwab to Glass, Grobman, and Moore, proposed public statement of the activities of BSCS, 17 February 1960, BSCS Papers.

39. BSCS Steering Committee meeting minutes, 2 February 1961, BSCS Papers.

40. BSCS Steering Committee meeting minutes, 13 March 1961, BSCS Papers.

41. Grobman to BSCS group, memo on controversial aspects of the curriculum, 8 June 1961, BSCS Papers.

42. See, for example, William Palmer Taylor and M. Phillips, "Dangers for Science? or, Snares for the Scientist?" *Science* 118 (1953): 449–450; Sidney Painter, H. A. Meyerhoff, and Alan T. Waterman, "The Visa Problem," *Scientific Monthly* 76 (1953): 11–19; Alan T. Waterman, J. Carlton Ward Jr., and Mervin J. Kelly, "Scientific Research and National Security," *Scientific Monthly* 78 (1954): 214–224; E. U. Condon, "The Duty of Dissent," *Science* 119 (1954): 227–228.

43. See Ellen W. Schrecker, *No Ivory Tower: McCarthyism and the Universities* (New York: Oxford University Press, 1986), 333; and Bentley Glass, "Academic Freedom and Tenure in the Quest of National Security," *BAS* 12 (1956): 221–226. On the antivivisectionist battle, see Glass, "Report on the Presidency of Bentley Glass, 1954–1956," unpublished typescript, not dated; and Glass to Cox, 26 August 1955, Glass Papers.

44. Glass, "Responsibilities of Biologists," 13.

45. Glass handwritten notes, 29 January 1960, BSCS Papers.

46. Grobman, memo on summer writing conference, 8 April 1960, BSCS Papers.

47. BSCS Steering Committee meeting minutes, 2 February 1961, BSCS Papers.

48. Patrick D. Wall to Zacharias, 8 November 1960, box 31, Killian Papers.

49. BSCS Steering Committee meeting minutes, 13 March 1961, BSCS Papers.

50. Moore to Grobman, 20 November 1961, BSCS Papers.

51. Grobman to BSCS Executive Committee, 14 March 1960; BSCS Steering Committee meeting minutes, 19 January 1962; BSCS Steering Committee meeting minutes, 7 December 1962; BSCS Steering Committee meeting minutes, 24 May 1963, BSCS Papers.

52. See Ronald L. Numbers, *The Creationists* (Berkeley: University of California Press, 1992), 238–239; and George E. Webb, *The Evolution Controversy in America* (Lexington: University Press of Kentucky, 1994), 130–142. Both of these works primarily discuss the reaction of religious groups to the reappearance of evolution in the high school textbooks. For a detailed look at the American reaction to Darwinism, particularly in its intersection with religion, see Numbers, *Darwinism Comes to America* (Cambridge: Harvard University Press, 1998).

53. Judith V. Grabiner and Peter D. Miller, "Effects of the Scopes Trial," *Science* 185 (1974): 832–837.

54. Vassiliki Betty Smocovitis, *Unifying Biology: The Evolutionary Synthesis and Evolutionary Biology* (Princeton: Princeton University Press, 1996), 134–138, 149.

55. Smocovitis, *Unifying Biology,* 171, 207.

56. Graduate School of Education, *Using Modern Knowledge to Teach Evolution in High School: As Seen by Participants in the High-School Conference of the Darwin Centennial Celebration at the University of Chicago, November 24–28, 1959* (Chicago: University of Chicago, Graduate School of Education, 1960).

57. Smocovitis, *Unifying Biology,* 156.

58. The major divisions of biology the Committee agreed on were: cell biology; organization and function of unicellular or noncellular units; organization and function of multicellular organisms; reproduction, growth and development; genetics; diversity of multicellular organisms; ecology; and evolution; BSCS Steering Committee meeting minutes, 5 February 1959, BSCS Papers.

59. Historian Walter P. Metzger even went so far as to attribute the modern notion of academic freedom to the ascendancy of the Darwinian world-view; see chapter 2 in *Academic Freedom in the Age of the University* (New York: Columbia University Press, 1955), 46–92.

60. Hermann J. Muller, "One Hundred Years Without Darwinism Are Enough," *School Science and Mathematics* 59 (1959): 310, 312.

61. On the growing influence of the fundamentalist movement in the 1950s, see Mark Silk, *Spiritual Politics: Religion and America Since World War II* (New York: Simon and Schuster, 1988), 53–64, 106–107; on the avoidance of controversy in textbooks, see Grabiner and Miller, "Effects of the Scopes Trial," 832–837.

62. John C. Mayfield, "Using Modern Knowledge to Teach Evolution in High School," in *Using Modern Knowledge,* Graduate School of Education, 11.

63. Riddle, "Amount and Nature of Biology Teaching in Secondary Schools," 69–75; Grabiner and Miller, "Effects of the Scopes Trial," 832–836.

64. Grobman to BSCS group, memo on controversial aspects of the curriculum, 8 June 1961, BSCS Papers.

65. Hermann J. Muller, "Man's Place in Living Nature," *Scientific Monthly* 84 (1957): 245. These thoughts were repeated in George Gaylord Simpson, "One Hundred Years Without Darwin Are Enough," *Teachers College Record* 62 (1961): 617–626. Simpson admittedly borrowed his title from the statement made earlier by Muller.

66. Muller, "One Hundred Years," 309.

67. BSCS Steering Committee meeting minutes, 2 February 1961, BSCS Papers.

68. The subject was prominent in the early outlines developed by the Committee on the Content of the Curriculum when it was thought that BSCS was going to produce only one textbook; BSCS Committee on the Content of the Curriculum meeting report, 30 October 1959, BSCS Papers.

69. BSCS Steering Committee meeting minutes, 2 February 1961, BSCS Papers

70. BSCS Steering Committee meeting minutes, 2 February 1961, BSCS Papers.

71. Muller, "One Hundred Years," 312.

72. For an example of each, see Moon, Mann, Otto, *Modern Biology,* 663–665; and William M. Smallwood, Ida L. Reveley, Guy A. Bailey, and Ruth A. Dodge, *Elements of Biology* (New York: Allyn and Bacon, 1948), 541–543.

73. Mayer to Grobman, 25 May 1960, box 9, Bates Papers.

74. George Gaylord Simpson, "The World into Which Darwin Led Us," *Science* 131 (1960): 969.

75. Grobman to BSCS group, memo on controversial aspects of the curriculum, 8 June 1961, BSCS Papers.

76. Ironically, the Scopes trial was resurrected during this era via *Inherit the Wind* not in response to public attitudes toward evolution, but rather as a scathing commentary on the political intolerance that swept the country during the Red Scare of the 1950s. The battle with the antievolutionists, many believed (though incorrectly), had already been fought and won in the 1920s; Edward J. Larson, *Summer for the Gods: The Scopes Trial and America's Continuing Debate Over Science and Religion* (New York: Basic Books, 1997), 239–246.

77. Glass to *New York Times,* 22 June 1960; BSCS Steering Committee meeting minutes, 9 June 1959, BSCS Papers.

78. Most members of the Steering Committee felt that topics related to health and hygiene should be moved into the earlier grades and that the new curriculum should move away from the past emphasis on human biology. At the same time, the Committee felt that man should be "brought into the picture more than is now the case in most high school textbooks." In this case, it was "man" as part of society, rather than "man" the individual; BSCS Steering Committee meeting minutes, 9 June 1959; Grobman, memo on summer writing conference, 8 April 1960, BSCS Papers.

79. Moore to Committee on the Content of the Curriculum, 16 July 1959, BSCS Papers.

80. Cox to Killian, 31 January 1958, box 5, OSAST.

81. AAAS, "Social Aspects of Science: Preliminary Report of AAAS Interim Committee," 146.

82. Taylor, "Is Biology Obsolete?" 14.

83. Grobman, "The Biological Sciences Curriculum Study," unpublished report, ca. April 1959, box 9, Bates Papers; see also Bentley Glass, "Renascent Biology: A Report on the AIBS Biological Sciences Curriculum Study," *School Review* 70 (1962): 94–119; and Glass to *New York Times,* 22 June 1960, BSCS Papers.

84. Glass handwritten notes, 28 January 1960, BSCS Papers.

85. Marston Bates, *The Forest and the Sea: A Look at the Economy of Nature and the Ecology of Man* (New York: Random House, 1960), 247. This book, written for a popular audience, parallels much of the context of the green version of the BSCS textbook; portions of it were listed as suggested readings at the end of many of the textbook's chapters.

86. BSCS, *High School Biology: BSCS Green Version* (Chicago: Rand McNally, 1963), 692–693. For the transformation of the population problem into a Cold War concern, see John Sharpless, "Population Science, Private Foundations, and Development Aid: The Transformation of Demographic Knowledge in the United States, 1945–1965,"in *International Development and the Social Sciences: Essays on the History and Politics of Knowledge,* ed. Frederick Cooper and Randall Packard (Berkeley: University of California Press, 1997), 176–200.

87. A summary of the Civil Rights movement can be found in James T. Patterson, *Grand Expectations: The United States, 1945–1974* (New York: Oxford University Press, 1996), 468–485.

88. BSCS Steering Committee meeting minutes, 7 December 1962, BSCS Papers.

89. BSCS, *Biological Science: An Inquiry into Life* [yellow Version] (New York: Harcourt, Brace & World, 1963), 671–672.

90. BSCS, *Biological Science: Molecules to Man* [blue Version] (Boston: Houghton Mifflin, 1963), 389.

91. Bentley Glass, "The Hazards of Atomic Radiations to Man—British and American Reports," *BAS* 12 (1956): 312–317; Robert A. Divine, *Blowing on the Wind: The Nuclear Test Ban Debate, 1954–1960* (New York: Oxford University Press, 1978), on Glass's activism, see 104, 135; on AEC minimization of fallout hazards, see 43, 116.

92. BSCS Steering Committee meeting minutes, 2 February 1961, BSCS Papers.

93. See BSCS Steering Committee meeting minutes, 9 June 1959; BSCS Steering Committee meeting minutes, 28 January 1960; Glass handwritten notes, 22 August 1960; BSCS Executive Committee meeting minutes, 22 November 1961, BSCS Papers. Despite his strong disagreement with the AEC over the danger of fallout, Glass urged BSCS to cooperate with the agency in the production of materials related to radiation biology. A flurry of atmospheric nuclear tests in late 1958 generated a "full-scale radiation scare" in 1959, see Divine, *Blowing on the Wind,* 262–280.

94. Bates, *Forest and the Sea,* 250.

95. Smocovitis, *Unifying Biology,* 148–149.

96. Simpson, "The World into Which Darwin Led Us," 973. Similar ideas were expressed by Muller and by Julian Huxley, "Evolution in the High School Curriculum," in *Using Modern Knowledge,* Graduate School of Education, 22–24. For a detailed description of the evolutionary movement in American thought, see Smocovitis, *Unifying Biology,* 138–153, 172–188.

97. *Zinjanthropus* is currently known as *Australopithecus boisei.*

98. AAAS, "Social Aspects of Science: Preliminary Report of AAAS Interim Committee," 146.

99. BSCS, *Biological Science: An Inquiry into Life,* 682–686.

100. Muller, "Man's Place in Living Nature," 253, 254.

101. BSCS, NSF Proposal, 29 April 1960, BSCS Papers.

102. BSCS, *Biological Science: An Inquiry into Life,* 686.

103. Schwab to Glass, Grobman, and Moore, proposed public statement of the activities of BSCS, 17 February 1960, BSCS Papers.

104. Joseph J. Schwab, *Biology Teachers' Handbook* (New York: John Wiley and Sons, 1963), 31, 39. Schwab's spelling of enquiry with an "e" was undertaken deliberately so that his work might not be mistaken for the work of educational psychologists who had begun to examine student problem solving under the heading of "inquiry;" see Ian Westbury and Neil J. Wilkof, introduction to *Science, Curriculum, and Liberal Education: Selected Essays of Joseph J. Schwab* (Chicago: University of Chicago Press, 1978), 3.

105. Muller, "Man's Place in Living Nature," 252.

106. Paulson to Richard T. Wareham, 24 October 1955, NAS/CEP.

107. BSCS Steering Committee meeting minutes, 13 May 1961, BSCS Papers.

108. Bentley Glass, "The Scientist's Perspective," *Science* 123 (1956): 777.

109. Moore, BSCS statement of objectives, 16 January 1960, BSCS Papers.

110. Much of Schwab's interest in this subject stemmed from his involvement in the development of the experimental undergraduate curriculum at the University of Chicago, a project initiated in the late 1920s by Robert M. Hutchins; Schwab, "The Science Programs in the College of the University of Chicago," in *Science and General Education,* ed. Earl McGrath (Dubuque: William C. Brown Co., 1947), 38–58; and "The Nature of Scientific Knowledge as Related to Liberal Education," *Journal of General Education* 3 (1949): 245–266.

111. Schwab, "The Teaching of Science as Inquiry," *BAS* 14 (1958): 374. This view of inquiry was spelled out clearly for teachers by Schwab in the *Biology Teachers' Handbook,* 39–41.

112. Schwab's academic writing included two articles on Dewey's work, "John Dewey: The Creature as Creative," *Journal of General Education* 7 (1953): 109–121; and "The 'Impossible' Role of the Teacher in Progressive Education," *School Review* 67 (1959):139–159. Westbury and Wilkof trace Schwab's intellectual debt to Dewey in the introduction to his collected essays, *Science, Curriculum, and Liberal Education,* 1–40.

113. Schwab, "Inquiry, the Science Teacher, and the Educator," *School Review* 68 (1960): 176. Other relevant essays include: Schwab, "Science and Civil Discourse: The Uses of Diversity," *Journal of General Education* 9 (1956): 132–143; and Schwab, "The Teaching of Science as Enquiry," in *The Teaching of Science,* Joseph J. Schwab and Paul F. Brandwein (Cambridge: Harvard University, 1960), 3–103.

114. Schwab, "Teaching of Science as Enquiry," 40.

115. Schwab, "Science and Civil Discourse," 139.

116. Schwab's views regarding the need for public support of scientific expertise were shared by most of the BSCS group. Muller, for example, wrote that "To give proper support to science, and to supply it with adequate

personnel, now requires every citizen to be taught to sympathize with the purposes of science," Hermann J. Muller, "Science for Humanity," *BAS* 15 (1959): 150.

117. Schwab, "Teaching of Science as Enquiry," 46–47.
118. Schwab, "Inquiry, the Science Teacher, and the Educator," 185.
119. Schwab, "Teaching of Science as Enquiry," 18, 38.
120. Schwab, "Inquiry, the Science Teacher, and the Educator," 184; Schwab, "Teaching of Science as Enquiry," 44.
121. Schwab, *Biology Teachers' Handbook,* 40–41.
122. Bert Kempers and Manert Kennedy, "Films and the Biological Sciences Curriculum Study," unpublished typescript, 1965; BSCS Film Committee meeting report, 12 December 1964, BSCS Papers.
123. BSCS Film Committee meeting report, 12 December 1964, BSCS Papers.
124. Bentley Glass quoted in Addison E. Lee, "The Block of Time Idea in Biology Laboratory Instruction," *American Biology Teacher* 22, no. 3 (1960): 135.
125. Glass to Dunham, September 1961, Glass Papers.
126. Lee, "Block of Time Idea," 136.
127. Glass in Lee, "Block of Time Idea," 136.
128. Schwab, *Biology Teachers' Handbook,* 46.
129. BSCS Steering Committee meeting minutes, 26 May 1962, BSCS Papers.
130. See Schwab, *Biology Teachers' Handbook,* 48, 53–54.
131. Schwab, *Biology Teachers' Handbook,* 46.
132. BSCS Steering Committee meeting minutes, 24 May 1963, BSCS Papers.

### Chapter 7

1. John Hersey, *The Child Buyer* (New York: Knopf, 1960), 33.
2. Hersey, *Child Buyer,* 34.
3. Paul Boyer, *By the Bomb's Early Light: American Thought and Culture at the Dawn of the Atomic Age* (Chapel Hill: University of North Carolina Press, 1994), 203–210; "The Critics Go to the Poll," *Saturday Review,* 1 October 1960, 17.
4. Frank G. Jennings, "Black Market in Brains," *Saturday Review,* 24 September 1960, 21; Sidney Shalett, "A Chapter of Our Time," *New York Times Book Review,* 25 September 1960, 4. Margaret Halsey, "The Shortest Way with Assenters," 21–22; B. F. Skinner, "May We Have a Positive Contribution," 22; Carl F. Hansen, "Educator vs. Educationist," 23; Robert Graham Davis, "An Arrangement in Black and White," 24; William Jay Smith, "The Truly Handicapped," 25–26, all in *New Republic,* 10 October 1960.
5. Halsey, "Shortest Way with Assenters," 22.
6. Hansen, "Educator vs. Educationist," 23.
7. "American Education and the Sciences," *New Republic,* 25 November 1957, 4–5; "Time and Money," *New Republic,* 6 January 1958, 7–8; Karl Shapiro, "Why Out-Russia Russia?," *New Republic,* 9 June 1958, 10; George Fischer, "Mistaken Envy: Soviet and American Education," *Progres-*

*sive,* March 1958, 14–22; Donald J. Hughs, "On Keeping Up with the Russians," *Nation,* 7 June 1958, 507–510; Eugene Lyons, "Soviet Education: Myth and Fact," *National Review,* 26 April 1958, 397–398.

8. Bestor to Hersey, 19 December 1960, box 98, Bestor Papers.

9. "Crisis in Education," *Life,* 24 March, 31 March, 7 April, 14 April, and 21 April, 1958; Hyman G. Rickover, *Education and Freedom* (New York: Dutton, 1959); "How Should We Retool Our Schools," *New York Times Book Review,* 1 February 1959, 1; Daniel Lee Kleinman and Mark Solovey, "Hot Science/Cold War: The National Science Foundation after World War Two," *Radical History Review* 63 (1995): 110–139.

10. David Tyack and Larry Cuban, *Tinkering Toward Utopia: A Century of Public School Reform* (Cambridge: Harvard University Press, 1995), 52; Joel Spring, *The Sorting Machine Revisited: American Educational Policy Since 1945,* updated ed. (New York: Longman, 1989), vii.

11. Scott L. Montogomery, *Minds for the Making: The Role of Science in American Education, 1750–1990* (New York: Guilford Press, 1994), 211.

12. Jerrold Zacharias, "A Scientist Looks at Scientists," in *The Scientific Revolution: Challenge and Promise,* ed. G. W. Elbers and P. Duncan (Washington, D.C.: Public Affairs Press, 1959), 138; Zacharias interview, PSSC/OHC.

13. Arnold Grobman, "The Biological Sciences Curriculum Study," unpublished report, ca. April 1959, box 9, Bates Papers.

14. Rockefeller Brothers Fund, *The Pursuit of Excellence: Education and the Future of America* (Garden City, NY: Doubleday, 1958), 28; Unedited transcript, Round Table No. 6, "Public Understanding of the Scientist," 3 February 1958, box 8, PCSE.

15. Warren Weaver, "A Great Age for Science," in *Goals for Americans,* President's Commission on National Goals (New York: The American Assembly, Columbia University, 1960), 105, 103.

16. Weaver, "Great Age for Science," 116.

17. Stuart W. Leslie, *The Cold War and American Science: The Military-Industrial-Academic Complex at MIT and Stanford* (New York: Columbia University Press, 1993); Thomas P. Hughes and Agatha C. Hughes, eds., *Systems, Experts, and Computers: The Systems Approach in Management and Engineering, World War II and After* (Cambridge: MIT Press, 2000); AAAS Committee on Science in the Promotion of Human Welfare, "Science and Human Welfare," *Science* 132 (1960): 69.

18. Joseph Turner, "When Boys Will Be Buys," *Science* 132 (1960):1367.

19. Harold Strauss to Hersey, 1 December 1960, box 286, Alfred A. Knopf, Inc. Records, Ransom Humanities Research Center, Austin, TX.

20. Joseph J. Schwab, "Science and Civil Discourse: The Uses of Diversity," *Journal of General Education* 9 (1956): 142.

21. Zacharias to various foundations, 18 June 1957, box 17, Zacharias Papers.

22. PSSC memo, "What We Seem to Know," 11 October 1956, box 1, PSSC/OHC.

23. Zacharias interview, PSSC/OHC.

24. C. Ray Carpenter handwritten notes, 9 September 1959, box 41, Carpenter Papers.

25. SPE Divisional Committee meeting minutes, 6 September 1956, NSF/HF.

26. Alan T. Waterman, "General Considerations Concerning United States Progress in Science and Education," unpublished typescript, 31 July 1958, box 4, OSAST.

27. Memo to the National Advertising Council, 24 October 1958; Eisenhower to Waterman, 28 August 1958; Waterman to Eisenhower, 31 July 1958, box 4, OSAST.

28. Ad Council Public Policy Committee meeting minutes, 20 November 1957, 20 November 1959, 18 November 1960, box 1, Ad Council Archives; ATW diary note, 13 July 1959, box 25, NSF/ODSF.

29. SPE Divisional Committee meeting minutes, 6 September 1956, NSF/HF. Waterman made this point in "Challenge to the American People," in *Scientific Revolution,* ed. Elbers and Duncan, 273.

30. C. Ray Carpenter handwritten notes, 9 September 1959, box 41, Carpenter Papers.

31. Unedited transcript, Round Table No. 6, "Public Understanding of the Scientist," 3 February 1958, box 8, PCSE.

32. MIT, "Project Troy Report to the Secretary of State," annex 14, p. 2, Troy Report Files. Concerns among liberal intellectuals over just this sort of domestic use of "propaganda" are discussed in Brett Gary, *The Nervous Liberals: Propaganda Anxieties from World War I to the Cold War* (New York: Columbia University Press, 1999).

33. Nathaniel Frank interview, PSSC/OHC; E. C. Wine to J. E. Webb, 8 May 1959, Webb Papers.

34. Zacharias to Dean McGee, 16 November 1956, box 16, Zacharias Papers.

35. Michels, "Teaching Aids-Texts," unpublished typescript, 1 January 1955, NAS/CPP.

36. Michels, "Teaching Aids-Texts," unpublished typescript, 1 January 1955, emphasis added NAS/CPP.

37. Zacharias to Killian, 15 March 1956, emphasis added, box 170, C/K Papers.

38. Stephen White, memo to the PSAC Education Panel, 11 August 1958, box 37, NSF/ODSF.

39. PSSC, "General Report," 25 March 1957, box 17, PSSC Papers.

40. PSSC, *Physics* (Boston: D.C. Heath, 1960); for emphasis on waves see Zacharias interview, PSSC/OHC.

41. Lab meeting minutes, May 20 1959; June 1 1959; June 10 1959, box 3; Guenter Schwarz, "Ripplers and Rippletanks; An Informal Report," June 8 1959, box 3, PSSC Papers.

42. Jerrold R. Zacharias, "Into the Laboratory," *Science Teacher* 24, no. 7 (1957): 324.

43. Friedman, "A Blueprint," *Science Teacher* 24, no. 7 (1957): 321.

44. Jerrold R. Zacharias, "Educational Methods and Today's Science—Tomorrow's Promise," *Technology Review* 59 (1957): 503.

45. Zacharias interview, PSSC/OHC.

46. PSSC, *Physics,* 381.

47. PSSC, *Physics,* 381.

48. Zacharias to Killian, 15 March 1956, box 170, C/K Papers.

49. Zacharias interview, PSSC/OHC; NSF proposal, 17 August 1956, box 17, Zacharias Papers. The move toward interdisciplinary work of this sort is described in Peter Galison, "The Americanization of Unity," *Daedalus* 127 (1998): 45–71.

50. Zacharias interview, PSSC/OHC.

51. Zacharias interview, PSSC/OHC.

52. Jack S. Goldstein, *A Different Sort of Time: The Life of Jerrold Zacharias* (Cambridge: MIT Press, 1992), 24–36.

53. Paul Forman, "Atomichron: The Atomic Clock from Concept to Commercial Product," *Proceedings of the IEEE* 73 (1985): 1181–1204.

54. Paul Forman, "Behind Quantum Electronics: National Security as Basis for Physical Research in the United States, 1940–1960," *HSPS* 18 (1987): 149–229; Bruce Hevly, "Tools of Science: Radio, Rockets, and the Science of Naval Warfare," in *National Military Establishments and the Advancement of Science and Technology,* ed. P. Forman and J. M. Sánchez-Ron (Dordrecht: Kluwer, 1996), 215–232; and Stuart W. Leslie, *Cold War and American Science.* There are a number of first-person appraisals of the influence of the war that run counter to this argument. Some physicists argued, for example, that the war provided them the opportunity to pursue research that was very much in line with their prewar work, see, for example, Jagdish Mehra and Kimball Milton, *Climbing the Mountain: The Scientific Biography of Julian Schwinger* (New York: Oxford University Press, 2000), 104–5. Such statements indicate the subtlety and complexity of the wartime influences on physics. My thanks to David Kaiser for this point.

55. Forman, "Behind Quantum Electronics," 190, 193.

56. "Hyphenated Scientists," *Nation,* 22 February 1958, 150.

57. The scientists at the time were certainly conscious of the potential for military distortion of their work, see Louis N. Ridenour, "Should the Scientists Resist Military Intrusion?," *American Scholar* 16 (Spring, 1947): 213–218, along with comments by Philip Morrison, Vannevar Bush, Norbert Wiener, and Donald Young, 218–223; also the follow up article, Albert Einstein, Alan T. Waterman, Robert K. Merton, William Y. Elliott, Douglas P. Adams, and Aldous Huxley, "Should the Scientists Resist Military Intrusion?: The Reader Replies," *American Scholar* 16 (Summer, 1947): 353–360; Paul Forman, "Into Quantum Electronics: The Maser as 'Gadget' of Cold War America," in *National Military Establishments,* ed. Forman and Sánchez-Ron, 261–326.

58. Roger L. Geiger, "Science and the University: Patterns from the US Experience in the Twentieth Century," in *Science in the Twentieth Century,* ed.

John Krige and Dominique Pestre (Amsterdam: Harwood Academic Publishers, 1997), 159–174.

59. On this point in particular, see Forman, "Behind Quantum Electronics," 229.

60. Killian quoted in Forman, "Behind Quantum Electronics," 201.

61. Goldstein, *Different Sort of Time*, 69–74.

62. Forman, "Behind Quantum Electronics," 216. This view is supported by the work of Ian Hacking, see, "Weapons Research," in Hacking, *The Social Construction of What?* (Cambridge: Harvard University Press, 1999), 163–185. For alternative views on the influence of military research expenditures on the form of scientific knowledge produced see Daniel Kevles, "Cold War and Hot Physics: Science, Security, and the American State, 1945–1956," *HSPS* 20 (1990): 239–264; and Roger Gieger, "Science, Universities, and National Defense, 1945–1970," *Osiris,* 2d series, 7 (1992): 26–48.

63. Forman, "Behind Quantum Electronics," 202–203.

64. On the effect of this shift on graduate and undergraduate education in physics see Leslie, *Cold War and American Science;* on the applied products generated from this work, see Robert Buderi, *The Invention that Changed the World* (New York: Simon and Schuster, 1996).

65. E. L. Ginzton, "Microwaves," *Science* 127 (1958): 841.

66. Ginzton, "Microwaves," 843; Forman, "Atomichron," 1200.

67. MIT, "Project Troy Report to the Secretary of State," 24, Troy Report Files.

68. Goldstein, *Different Sort of Time,* 120–121.

69. Evelyn Fox Keller, "Physics and the Emergence of Molecular Biology: A History of Cognitive and Political Synergy," *JHB* 23 (1990): 389–409; Lily E. Kay, "Problematizing Basic Research in Molecular Biology," in *Private Science: Biotechnology and the Rise of the Molecular Sciences,* ed. Arnold Thackray (Philadelphia: University of Pennsylvania Press, 1998), 20–38; Pnina G. Abir-Am, "The Molecular Transformation of Twentieth-Century Biology," in *Science in the Twentieth Century,* ed. Krige and Pestre, 495–524. Much of this transformation in biology was underway in the 1930s as a result of the program in molecular biology established by the Rockefeller Foundation. World War II, however, accelerated the pace of the transition considerably.

70. On the social uses to which biology was often put in schools, see Philip J. Pauly, *Biologists and the Promise of American Life: From Meriwether Lewis to Alfred Kinsey* (Princeton: Princeton University Press, 2000), 171–193.

71. Committee on Educational Policies, *Recommendations on Undergraduate Curricula in the Biological Sciences* (Washington, D.C.: National Research Council, 1958), 23.

72. Truman J. Moon, Paul B. Mann, and James H. Otto, *Modern Biology,* rev. ed. (New York: Henry Holt and Company, 1951). Another commonly used textbook of the time was Ella Thea Smith, *Exploring Biology,* 5th ed. (New York: Harcourt Brace, 1959). Both these texts were viewed by BSCS Steer-

ing Committee members as representative of the current state of the biology curriculum, Grobman to Content Committee, 9 February 1960, box
9, Bates Papers.

73. Harry J. Fuller, "The Emperor's New Clothes, or *Prius Dementat,*" *Scientific
Monthly* 72 (1951): 38.

74. Nicholas Rasmussen, "The Mid-Century Biophysics Bubble: Hiroshima
and the Biological Revolution in America, Revisited," *History of Science* 35
(1997): 245–293; and Lily E. Kay, "Conceptual Models and Analytical
Tools: The Biology of Physicist Max Delbrück," *JHB* 18 (1985):207–246.

75. Keller, "Physics and the Emergence of Molecular Biology," 392.

76. Wallace O. Fenn, "Front Seats for Biologists," speech given before the AIBS
annual meeting, Stillwater, OK, 29 August 1960, box 3, Wallace O. Fenn
Papers, Edward G. Miner Library, University of Rochester Medical Center, Rochester, NY.

77. Secondary School Biological Sciences Film Series meeting minutes, 8–9
October 1958, box 2, Sonneborn Papers.

78. Moore, BSCS: Penultimate Revision of Objectives, 16 January 1960, BSCS
Papers.

79. Schwab to Steering Committee, 1 June 1959; Moore to Content Committee, 16 July 1959; BSCS Steering Committee meeting minutes, 9 June
1959, BSCS Papers.

80. Krutch to Grobman, 12 September 1959, BSCS Papers.

81. Prosser to Grobman, excerpted in BSCS memo No. 28, 26 October 1959,
box 9, Bates Papers.

82. AIBS Long Range Planning Council meeting minutes, 12 December
1961, box 3, Muller Papers.

83. Toby A. Appel, *Shaping Biology: The National Science Foundation and American Biological Research, 1945–1975* (Baltimore: The Johns Hopkins University Press, 2000), 208.

84. NSF staff paper, "Perspectives in the Life Sciences," ca. May 1963,
NSF/HF.

85. NSF staff paper, "Perspectives in the Life Sciences," ca. May 1963,
NSF/HF.

86. Grobman to Steering Committee, 6 December 1962, box 3, Muller Papers.

87. Grobman to Steering Committee, 6 December 1962, 30 January 1963, box
3, Muller Papers.

88. Grobman to Steering Committee, 6 December 1962, 30 January 1963, box
3, Muller Papers.

89. Grobman to Steering Committee, 6 December 1962, box 3, Muller Papers.

90. Grobman to Executive Committee, 14 March 1960, BSCS Papers.

91. BSCS supplementary film list, 3 November 1962, box 10, Bates Papers.

92. Glass, unpublished typescript on laboratory innovations, ca. January 1960,
BSCS Papers; Addison E. Lee, David L. Lehman, and Glen E. Peterson,
eds., *Laboratory Blocks in Teaching Biology,* BSCS Special Publication no. 5

(Washington, D.C.: American Institute of Biological Sciences, 1967), 8; Addison E. Lee, *Plant Growth and Development: A Laboratory Block* (Boston: D.C. Heath, 1963); Florence Moog, *Animal Growth and Development: A Laboratory Block* (Boston: D.C. Heath, 1963). The connection between plant hormone research and World War II is made in Nicholas Rasmussen, "Plant Hormones in War and Peace: Science, Industry, and Government in the Development of Herbicides in 1940s America," *Isis* 92 (2001): 291–316.

93. BSCS Steering Committee meeting minutes, 26 May 1962, BSCS Papers; Joseph J. Schwab, *Biology Teachers' Handbook* (New York: John Wiley and Sons, 1963).

94. Appel, *Shaping Biology,* 216–222.

95. See, for example, David P. Ausubel, "An Evaluation of the BSCS Approach to High School Biology," *American Biology Teacher* 28 (1966): 176–184; G. L. Stebbins to Grobman, n.d., box 9, Bates Papers.

96. Marston Bates, *The Natural History of Mosquitoes* (New York: Macmillan, 1949) and *The Nature of Natural History* (New York: Charles Scribner's Sons, 1950).

97. Bates, *Nature of Natural History,* 276, 278.

98. Joel B. Hagen, *An Entangled Bank: The Origins of Ecosystem Ecology* (New Brunswick, NJ: Rutgers University Press, 1992), 100–121.

99. Hagen, *Entangled Bank,* 78–145; See also, Peter J. Taylor, "Technocratic Optimism, H. T. Odum, and the Partial Transformation of Ecological Metaphor after World War II," *JHB* 21 (1988): 213–244; Stephen Bocking, "Ecosystems, Ecologists, and the Atom: Environmental Research at Oak Ridge National Laboratory," *JHB* 28 (1995): 1–47

100. BSCS, *High School Biology: BSCS Green Version* (Chicago: Rand McNally, 1963), 1–75.

101. Vassiliki Betty Smocovitis, *Unifying Biology: The Evolutionary Synthesis and Evolutionary Biology* (Princeton: Princeton University Press, 1996), 134–138, 149.

102. Muller, "Remarks Concerning the Content of a High School Biology Course," unpublished typescript, box 3, Muller Papers.

103. NSF staff paper, "Perspectives in the Life Sciences," ca. May 1963, NSF/HF.

104. Glass to Dunham, ca. September 1961; Addison Lee to Glass with AEC lab block proposal, 25 September 1961, Glass Papers.

105. Grobman to Glass, 11 September 1961, Glass Papers.

106. Galison, "Americanization of Unity," 45–71; Andy Pickering, "Cyborg History and the World War II Regime," *Perspectives on Science* 3 (1995), 6–33.

## Conclusion

1. Joseph J. Schwab, *Biology Teachers' Handbook* (New York: John Wiley and Sons, 1963), 3–8.

2. Joseph J. Schwab, "The Teaching of Science as Enquiry," in *The Teaching of Science,* Joseph J. Schwab and Paul F. Brandwein (Cambridge: Harvard University Press, 1962), 3.

3. Examples of the conventional interpretation are found in Diane Ravitch, *The Troubled Crusade: American Education, 1945–1980* (New York: Basic Books, 1983), 228–230; Joel Spring, *The Sorting Machine Revisited: American Educational Policy Since 1945,* updated ed. (New York: Longman, 1989), 93–150; and George E. DeBoer, *A History of Ideas in Science Education: Implications for Practice* (New York: Teachers College Press, 1991), 147–157.

4. Daniel S. Greenberg, *The Politics of Pure Science* (New York: New American Library, 1967), 51–67; For a more comprehensive examination of the relationship between science and the federal government prior to World War II, particularly the government's involvment in agriculture, see A. Hunter Dupree, *Science in the Federal Government: A History of Policies and Activities to 1940* (Cambridge: Harvard University Press, 1957).

5. Kenneth M. Jones, "The Government-Science Complex," in *Reshaping America: Society and Institutions, 1945–1960,* ed. Robert H. Bremner and Gary W. Reichard (Columbus: Ohio State University Press, 1982), 315–342.

6. These concerns are well documented in the AAAS report, "Social Aspects of Science: Preliminary Report of AAAS Interim Committee," *Science* 125 (1957): 143–147; AAAS, "1958 Parliament of Science," *Science* 127 (1958): 852–858; and Don K. Price, J. Stefan Dupré, and W. Eric Gustafson, "Current Trends in Science Policy in the United States," *Impact of Science on Society* 10 (1960): 187–213. The scientists' appreciation of their entanglement in national affairs can be seen in Philip Morrison, "Atoms and the Enlargement of Science," *Journal of General Education* 1 (1947): 211–215; James B. Conant, "Science and Politics in the Twentieth Century," *Foreign Affairs* 28 (1950): 189–202; Bentley Glass, "The Academic Scientist, 1940–1960," *Science* 132 (1960): 598–603; and Glass, "Scientists in Politics," *BAS* 18, no. 5 (1962): 2–7; among many others.

7. AAAS, "Parliament of Science," 853.

8. Steven Shapin describes institutional trust as a defining feature of modern-day knowledge societies. He notes the explicit recognition of this by sociologist Robert K. Merton in the early 1940s, which coincided with the emergence of the expert society in the middle of the twentieth century; Shapin, *A Social History of Truth: Civility and Science in Seventeenth-Century England* (Chicago: University of Chicago Press, 1994), 409–417.

9. Morrison, "Enlargement of Science," 213, 215.

10. See, for example, Herbert M. Kliebard, *The Struggle for the American Curriculum: 1893–1958,* 2d ed. (New York: Routledge, 1995), 77–105.

11. These models along with their underlying assumptions have exerted a powerful influence in the decades since, and, in many ways, underlie current, NSF-supported systemic reform efforts in science education. An understanding of the origins and original purposes of this approach, at the very least allows us to see our current efforts in a different light. For an elaboration of this point, see John L. Rudolph, "From World War to Woods Hole: The Use of Wartime Research Models for Curriculum Reform," *Teachers College Record* 104 (2002): 212–241.

# INDEX